Green Plant Extract-Based Synthesis of Multifunctional Nanoparticles and their Biological Activities

Authored By

Seyed Morteza Naghib

School of Advanced Technologies
Iran University of Science and Technology
Tehran, Iran

&

Hamid Reza Garshasbi

School of Advanced Technologies
Iran University of Science and Technology
Tehran, Iran

Green Plant Extract-Based Synthesis of Multifunctional Nanoparticles and their Biological Activities

Authors: Seyed Morteza Naghib & Hamid Reza Garshasbi

ISBN (Online): 978-981-5179-15-6

ISBN (Print): 978-981-5179-16-3

ISBN (Paperback): 978-981-5179-17-0

need for a court order if at any point you breach any terms of this License Agreement. In no event will any delay or failure by Bentham Science Publishers in enforcing your compliance with this License Agreement constitute a waiver of any of its rights.

3. You acknowledge that you have read this License Agreement, and agree to be bound by its terms and conditions. To the extent that any other terms and conditions presented on any website of Bentham Science Publishers conflict with, or are inconsistent with, the terms and conditions set out in this License Agreement, you acknowledge that the terms and conditions set out in this License Agreement shall prevail.

Bentham Science Publishers Pte. Ltd.
80 Robinson Road #02-00
Singapore 068898
Singapore
Email: subscriptions@benthamscience.net

BENTHAM SCIENCE

CONTENTS

PREFACE

Nanobiotechnology is gaining tremendous impetus in this era owing to its ability to modulate metals into their nano size, which efficiently changes their chemical, physical, and optical properties. Accordingly, considerable attention is being given to the development of novel strategies for the synthesis of different kinds of nanoparticles of specific composition and size using biological sources. However, most of the currently available techniques are expensive, environmentally harmful, and inefficient with respect to materials and energy use. Several factors, such as the method used for synthesis, pH, temperature, pressure, time, particle size, pore size, environment, and proximity, greatly influence the quality and quantity of the synthesized nanoparticles and their characterization and applications. In recent years, developing efficient green chemistry methods for synthesizing metal nanoparticles has become a major focus of researchers. They have investigated in order to find an eco-friendly technique for the production of well-characterized nanoparticles. One of the most considered methods is the production of metal nanoparticles using organisms. Among these organisms, plants seem to be the best candidates, and they are suitable for large-scale biosynthesis of nanoparticles. Nanoparticles produced by plants are more stable, and the synthesis rate is faster than in the case of microorganisms.

Moreover, the nanoparticles are more varied in shape and size than those produced by other organisms. The advantages of using plant and plant-derived materials for the biosynthesis of metal nanoparticles have interested researchers in investigating mechanisms of metal ions uptake and bio-reduction by plants and understanding the possible mechanism of metal nanoparticle formation in plants. In this review, most of the plants used in metal nanoparticle synthesis are shown.

Seyed Morteza Naghib
School of Advanced Technologies
Iran University of Science and Technology
Tehran, Iran

&

Hamid Reza Garshasbi
School of Advanced Technologies
Iran University of Science and Technology
Tehran, Iran

CHAPTER 1

Green Synthesis and Antibacterial Activity of Noble Metal Nanoparticles using Plants

Abstract: The emerging properties of noble metal nanoparticles (NPs) are attracting huge interest from the translational scientific community and have led to an unprecedented expansion of research and exploration of applications in biotechnology and biomedicine. An array of physical, chemical and biological methods has been used to synthesize nanomaterials. In order to synthesize noble metal NPs of particular shapes and sizes, specific methodologies have been formulated. Although ultraviolet irradiation, aerosol technologies, lithography, laser ablation, ultrasonic fields, and photochemical reduction techniques have been used successfully to produce NPs, they remain expensive and involve hazardous chemicals. Therefore, there is a growing concern about developing environment-friendly and sustainable methods. Since the synthesis of nanoparticles of different compositions, sizes, shapes and controlled dispersity is an important aspect of nanotechnology, new cost-effective procedures are being developed. Microbial synthesis of NPs is a green chemistry approach that interconnects nanotechnology and microbial biotechnology. Biosynthesis of gold, silver, gold–silver alloy, selenium, tellurium, platinum, palladium, silica, titania, zirconia, quantum dots, magnetite, and uraninite nanoparticles by bacteria, actinomycetes, fungi, yeasts, and viruses have been reported. However, despite stability, biological NPs are not monodispersed, and the rate of synthesis is slow. To overcome these problems, several factors, such as microbial cultivation methods and extraction techniques, have to be optimized, and the combinatorial approach, such as photobiological methods, may be used. Cellular, biochemical and molecular mechanisms that mediate the synthesis of biological NPs should be studied in detail to increase the rate of synthesis and improve the properties of NPs.

Keywords: Nanoparticles, NMPS, Nanotechnology, Nanoparticles synthesis.

INTRODUCTION

Metal NPs have a lengthy preparation, characterization, and use history in several fields. Nanomaterials research is spurred by the desire to understand better the characteristics of noble metal nanoparticles (NMNPs) and to discover how they might be employed in various applications. The high surface-to-volume ratio of NPs, as well as the confinement of electrons, phonons, and electric fields, confer a broad variety of properties on them. NPs have a high surface-to-volume ratio,

Seyed Morteza Naghib and Hamid Reza Garshasbi

which is partially responsible for this. Because of its high surface energy and substantial curvature, the nanoparticle's surface may become unstable. NPs surfaces have a particularly high proportion of curved regions in the form of edges and corners. Edges and corners are more likely to have hanging bonds, *i.e.,* coordinately unsaturated atoms, than flat surfaces. On the surface, corners, and edges of NPs, there are many atoms with uncoordinated atoms that affect the particle's chemical reactivity and surface bonding. By changing quantum levels and altering transition probabilities, the electron confinement effect in NPs alters the spectrum features of the particle. Particle-particle and particle-environment interactions, as well as volume ratio and confinement phenomena, are influenced by the large surface area. In recent years, scientists have learned how NPs shape affects their properties. Because of their metastable properties, NPs with non-spherical geometries are effectively locked in motion. Morphology controls further alterations in internal structures, surface properties, and orientational confinement [1].

NOBLE METAL NANOPARTICLES (NMNPS)

According to their size and usual structure, NPs may have various characteristics. It is possible to employ NPs in novel ways because of their high surface-to-volume ratio.

Additionally, there is an increase in unsaturated bonds, as well as a shift in bandgap energies. In order to make nanomaterials for particular purposes, NPs must be synthesized under strict supervision. These advances allow for the development of nanostructures with specific topological and morphological attributes and specific functional properties. Metallic NPs, polymeric NPs, and magnetic NPs are plentiful. The hydrophilicity or hydrophobicity of NPs and their functionalization greatly impact their practicality. It is possible to use NPs in various applications, such as nanomedicine, drug delivery, sensors, and optoelectronics. The unique physical-chemical features of noble metal NPs (NMNPS) make them extremely versatile. AuNPs, AgNPs, and PtNPs are stable noble metal NPs materials that may be easily synthesized chemically and customized in surface functionalization. NMNPs identify bioactive compounds and pollutants using colorimetry, immunoassays, Raman spectroscopy, and sensors. This paper examines the growing use of noble metallic nanoparticles in food safety. Bioactive compounds and trace pollutants are highlighted in this chapter [2].

METHODOLOGIES OF NPS SYNTHESIS

Chemical Methods for the Synthesis of NPs

Chemical Reduction

Colloidal metal particles can be made using a simple chemical reduction process that does not require expensive equipment. Chemical-reducing agents such as sodium borohydride and citrate are the most commonly used agents. Smaller NPs are produced by powerful reducing agents than by weak reducing agents. Oligomers, clusters, and precipitates are generated from excess surface energy and thermodynamic instability in smaller particles [3].

Co-precipitation

In order to make MNPs, co-precipitation is a simple and effective process. Since 1981, when Massart reported on the creation of MNPs under acid and alkaline conditions, iron oxide MNPs have been made in this manner. To reduce a metallic ion (*e.g.,* Fe^{2+} and Fe^{3+}) combination, a basic solution (typically $NaOH$, NH_3OH, or $N(CH_3)_4OH$) is used in the following chemical process at temperatures below 100°C.

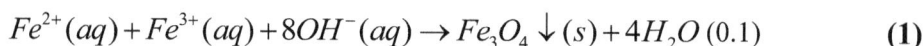

$$Fe^{2+}(aq) + Fe^{3+}(aq) + 8OH^-(aq) \rightarrow Fe_3O_4 \downarrow (s) + 4H_2O \ (0.1) \tag{1}$$

Since organic solvents are not required, the co-precipitation process is simple to repeat and inexpensive. However, reaction conditions significantly impact the particle size, shape, and content. Molecularly, light surfactants or functionalized polymers are required for stabilization. To make matters worse, iron oxide particles created in this method are typically unstable.

Sol-gel

Metal alkoxides or their precursors are often used in the condensation and hydrolysis reaction of sol-gel techniques to produce NPs. The intermediates must be heated to achieve good crystallinity in the produced NPs. Precursors for forming oxide particles that interact by van der Waals forces or H-bonding and are dispersed in a "sol" gelled by solvent evaporation or other chemical processes are metal alkoxides, which are used in this procedure. However, the alkoxide precursors are hydrolyzed in a base or acid. This results either in a colloidal gel or a polymeric gel, which can be used as a solvent in general. Condensation and hydrolysis rates greatly impact the final product's properties. Slower hydrolysis yields smaller NPs.

Magnetic ordering substantially affects the dispersion, formation of phases, volume fraction, and size distribution in a sol-gel system. The main disadvantage of the sol-gel approach is that it introduces contaminants from reaction byproducts, necessitating subsequent treatment [4].

Microemulsion

This kind of dispersion is known as microemulsions (MEs) because it is monophasic; optically isotropic; thermodynamically stable, transparent, and monochromatic. The reflected light from certain microemulsions is white, but the transmitted light is often reddish. Oil, water, and a surfactant mixture are the most common components of microemulsions. A zwitterionic layer may divide water-rich and oil-rich areas in microemulsions. Hoar and Schulman (1943) argued that the microemulsion's properties are uncertain. There is not a clear statement of the many stages and structures involved. Water or oil droplets and bicontinuous structures may form within the microemulsion domains [5]. Because of the surfactant and polar phase (usually water), a thermodynamically stable and homogenous microemulsion is created by mixing water with oil and a surfactant. Surfactant molecules produce an interfacial layer microscopically dividing the polar and nonpolar domains. Microstructures, ranging from oil droplets in a continuous water phase to water droplets in a continuous oil phase, can be found in this interfacial layer. As nanoreactors, the latter creates NPs with minimum polydispersity. For example, many microemulsions, such as sc-CO_2 (w/sc-CO_2), have been found [6].

Hydrothermal

Powders produced by hydrothermal synthesis have superior features such as purity, phase stability, and controllable morphologies based on advancements in particle technology. At high temperatures (T >251°C) and high pressures (>100 KPa), crystals are formed directly from solutions in aqueous environments. Particle sizes and morphologies can be precisely regulated, and aggregation/aggregation is minimized by using this technique [7].

Solvothermal

Solution-based nanowire growth using solvothermal nanowire synthesis is a catalyst-free, high-pressure technique. Chemical reactions are carried out in a high-pressure reactor using an organic solution containing semiconductor precursors and surfactants at the boiling point of the solvent. Partial evaporation may induce hundreds of bars increases in liquid phase pressure, encouraging precursor breakdown and crystal nucleation. It is also possible to achieve substantial growth rates by depositing surfactant on the nanowire sidewalls to avoid agglomeration. A hydrothermal synthesis procedure uses water as the solvent. Because it does not use a hazardous or flammable solvent, this method has higher temperatures and pressures. This is known as hydrothermal synthesis when water is used as the solvent.

For this reason, it may be used at greater temperatures and pressures since there is no dangerous or flammable solvent. Large-scale production is made possible by its ease of use and low-temperature budget. Only thermoelectric materials with highly anisotropic crystal structures and preferred growth orientations, such as PbTe or SbTe, or Bi2Te$_3$, may benefit from this technique [8].

Sonochemical Synthesis

Ultrasound irradiation has been used in electrochemistry operations since the 1930s. However, in the last ten years, sonoelectrochemistry has grown in importance. Microbubbles in the electrolyte may be linked to a wide spectrum of ultrasonic waves' effects on electrochemistry processes. If cavitation occurs near the electrode's surface, what should be done? In this example, a high-velocity liquid microjet travels parallel to the electrode in the direction of the electrode surface. At higher than ultrasonic threshold intensity, shock waves and microstreaming can also cause a bubble's collapse. The diffusion layer's thickness decreases due to all of these events. It is possible to improve the overall flow of mass and reaction speeds and clean and degas the electrode's surface. There were additional chemical consequences related to radical production from solvent sonolysis. Environmental cleanup and nano powder production have recently increased interest in sonoelectrochemistry.

The introduction of ultrasonic irradiation into electrochemical systems has been accomplished in various ways. Sonoelectrochemistry cells have been studied using ultrasound probes. An ultrasonic bath was submerged in a typical electrochemistry cell in a fixed place as the first and the most basic setup. In several investigations, this setup was used. However, because the ultrasonic field is not uniformly distributed, the power conveyed in the electrochemistry cell is limited. The outcomes are heavily influenced by where it is placed. If you prefer another method, you may put an ultrasonic probe or horn system straight into an electrochemistry cell. By aiming the ultrasonic waves at the electrode surface, more accurate power management is made feasible.

They have positioned face-to-face and a certain distance apart in the solution. An alternative is to use the ultrasonic horn as the actual electrode. So-called "sonotrodes" or "solo electric electrodes," on the other hand, refer to these devices.

There have been several attempts to study the electrodeposition of copper using a new, novel type of sonoreactor. However, Reisse and colleagues were the first to employ this new method. A sonotrode system utilizing electrolysis and ultrasonic pulses sequentially was employed to generate nano powders. It has also been used to study the electroreduction of benzaldehyde and benzoquinone.

Reisse *et al.* present a pulsed sonoelectrochemical reduction system to synthesize metal powders. Fig. (**1**) depicts the experimental setup. The titanium probe (20 kHz) was used as a cathode and an ultrasonic emitter in these studies. An isolating plastic jacket covers the cylindrical portion of the sonoelectrode immersed in the electrolyte at the horn's bottom. A pulse driver connects an ultrasound probe to a generator and potentiostat. Galvanostatic action requires a two-electrode cell in the original design. Using galvanostatic control to prevent undesirable secondary reactions is a downside of this design. An adaptation was made to counteract this. As a result, a three-electrode configuration, rather than a two-electrode configuration, was implemented in the sonoelectrochemistry system. There are many reasons why galvanostatic conditions have been used for most of these processes. They are less complicated and can be utilized to mass-produce many NPs.

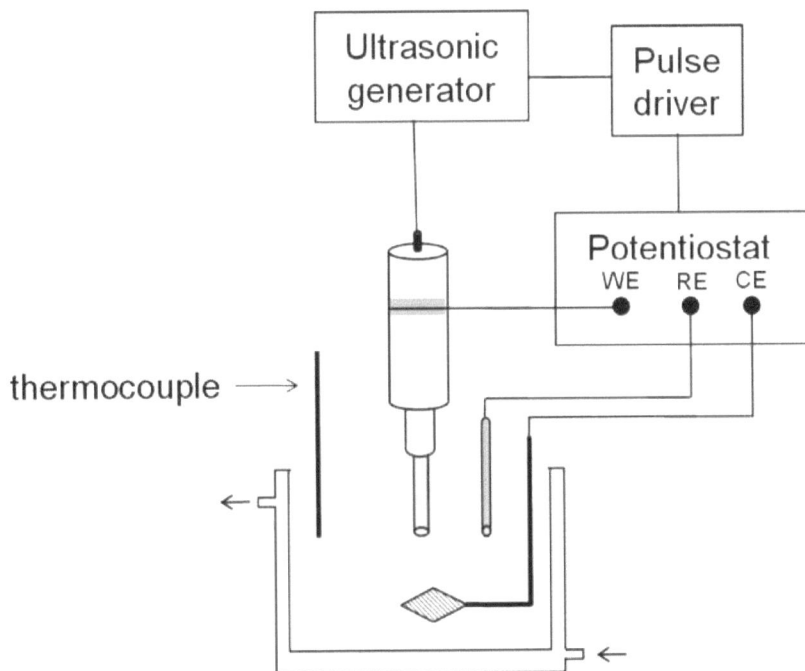

Fig. (**1**). Schematic of Sonoelectrochemistry setup [9].

Before doing any sonoelectrochemistry experiment, it is critical to determine the ultrasonic power given to the cell. Nucleation is at the foundation of pulsed sonoelectrochemical production of nano powders, which uses massive nucleation. At cathode, metal nuclei on the surface of the sonoelectrode are decreased, resulting in a thick covering of metal nuclei. The titanium horn is employed as an electrode (TON). Metal particles on the cathode surface are removed by a brief TUS pulse. By whirling the solution, it supplies the two layers with metal cations. To help reestablish the initial conditions at the sonoelectrode surface, a rest time (TOFF) after the two previous pulses may be helpful.

Ultrasonic horns are made using this titanium alloy. An oxide layer covers the titanium's surface, consisting of TiO_2, Ti_2O_3, and oxygen absorbed from the air. A passivated layer applied to the sonoelectrode surface during an oxidation process can provide insulation. The employment of a sonoelectrode in the reduction process is restricted by this restriction. It is important to polish the titanium sonoelectrode before each experiment to remove any contaminants that may interfere with the nucleation process. Numerous nano powders of pure metals or alloys and semiconductor NPs have been made using this unique electrochemical approach. Pulsed sonoelectrochemistry has also resulted in the creation of conductive polymer NPs. Finely split metal powders have a particle size of 100 nm, a huge surface area, and are chemically pure [9].

Microwave Synthesis

As a result of this interaction, microwave-aided syntheses may be used to synthesize nonpolar solvent molecules. Electromagnetic waves can directly interact with solution/reactants with excellent energy efficiency and reduced synthesis time to create fast and homogeneous heating. MW heating can yield smaller crystals because local superheating leads to the rapid growth of many seeds.

As a result, microwave (MW) and synthetic ultrasonic (SUS) technologies have become more popular due to their ability to produce high-quality goods in short periods. MW and US techniques provide fast crystallization, homogeneous nucleation, simple morphological control, phase selectivity, particle size reduction, and quick warming. The ability to control particle size distribution is another advantage of MW, as smaller distributions are more common with faster reaction times. Previously, microwave irradiation was used to make larger quantities of MOF-5. Variables such as microwave power, radiation time, temperature, solvent concentration, and substrate composition impacted the final product's crystallinity and shape. Microwave power, irradiation time, temperature, solvent concentration, and substrate type all impacted the product's crystallinity and form. The MW irradiation time and power level are crucial to bear in mind while synthesizing MOFs with smaller dimensions. When the MW irradiation time and power increase, bigger crystals (between 20 and 25 nm) may be produced [10].

Spray Pyrolysis

Aerosols are produced using spray pyrolysis from various precursor solutions, such as metallic salts or colloidal solutions. When the solution droplets have been heated to a predetermined temperature, the solvent is evaporated from the surface, the droplets are dried, and the precipitated solute is dried. High temperatures anneal the precipitate, producing microporous particles with a predetermined phase composition. Solid particles are created, and sintering is completed. Sintering "*in situ*" is required due to the highly reactive character of the particles formed during thermolysis. Spray pyrolysis calls for uniform and fine droplets of reactants to be prepared, as well as a controlled thermal degradation of those droplets. Other synthetic methods cannot compete with spray pyrolysis in terms of benefits. The spray pyrolysis process uses inexpensive equipment and an experimental setup to save money.

It is also not necessary to utilize high-quality reagents and formulations. Particle shape and size may be fine-tuned further by adjusting the preparative conditions, including additives, flow rate, and reaction concentration. Continuous methods are also available for morphological control and the creation of fine powders with round particles and the required diameter dictated by the droplet size. Some downsides of spray pyrolysis include: (1) it is difficult to scale up (yields are small), (2) oxidation of sulfur compounds can occur when treated in an air atmosphere, and (3) it is difficult to determine the growth temperature. Low-cost spray pyrolysis can provide high-density packaging and particle uniformity in porous films and films. Powders with tiny particle sizes (less than 1 mm), narrow size distribution (1- 2 mm), excellent purity, and significant surface area may be created utilizing this manufacturing method. System components used in spray pyrolysis include atomizers, precursor solutions, substrate heaters, and temperature controls. Spray pyrolysis has been used to manufacture a wide range of thin films, including solar cells, sensors, and solid oxide fuel cells [11].

Laser Pyrolysis

Continuous-wave CO_2 lasers are used to heat gases, triggering molecular decomposition and the production of NPs in a process known as Laser Pyrolysis. Molecular decomposition occurs when the laser beams cross precursor gases, which absorb the laser energy (nuclei). An inert gas moves the NPs to a collection bag when nuclei have reached the critical stage of homogenous nucleation.

Coalescence is more intense at high temperatures, resulting in spherical particles, but different shapes can be formed at low temperatures. Process variables may be used to alter the characteristics of NPs. This process yields fine, uniform particles of excellent quality. The approach yields high-purity nanomaterials continuously due to the small number of side reactions [12].

Wet Chemical Etching

The procedure of glass microfabrication that sees the greatest amount of application is called wet chemical etching. Hydrofluoric acid is the most common

etchant for silicate glass (HF). Other ingredients like HCl, HNO_3, and NH_4F-buffer can be used. The etching chemical reaction is depicted below:

$$SiO_2 + 6HF \rightarrow H_2SiF_6 + 2H_2O \qquad (2)$$

Wet chemical etching results in microchannels with rounded sidewalls and isotropy. During the wet etching process, the use of titanium as a receding mask enables the form and angle of the sidewall to be altered. The etch rate and the etch time define the channel depth. If the mask opening is multiplied by two, the channel's width may be determined. Because of its static nature, gold (Au) is a good masking material for HF etching. Glass wet etching masks are often made of chromium (Cr), gold, or a thin film layer. Photolithography is another common masking method. Plasma dry etching creates an Au Poly-Si layer that is adhered to the glass with the help of Cr. An inexpensive, low-cost masking material may be a thick negative photoresist coating (such as the popular SU-8) for some shallow etchings. Crystalline quartz can be used as a substrate for anisotropic wet glass etching. As a result, Z-cut wafers are the most popular choice since they have the greatest etch rate of any other kind. Microfluidic structures may be built and modified using etching settings [13].

Electro-explosion

Electrical energy is stored for a long time before being released in a burst using high-power pulse technology. High voltage, high current, and a powerful pulsed discharge are created in the process.

Marx energy storage module, bipolar charge power supply, a high-voltage pulse trigger, wire, and discharge protection switch are used to replicate an electric explosion in this experiment.

To begin with, a bipolar charge power source is used to recharge the Marx energy storage module. An electric explosion source's three-electrode switch is activated by a signal sent by the control system when charging is complete. The circuit for discharging waste is now active. The wire undergoes a transformation from a solid to a plasma state. The voltage and current supplied to it cause it to expand fast. Specimen fragmentation occurs due to the wire explosion converting electrical energy into shock wave energy. There is a total capacity of 4 F, a charging voltage of 60 kV, and an energy capacity of 7.2 kJ in the electric explosion system [14].

Thermal Decomposition

Temperature, reactant concentrations, stabilization agents (surfactants), and surfactants all play a role in establishing a controlled nanometric size for a given reaction time. According to Palacious-Hernandez and Kino's research, the solventless thermal decomposition process is a simple and moderate route that requires no raw materials. The biological method generated just a small number of NPs, on the other hand, as noted by Tran *et al.*, heat breakdown produced far more NPs. The injection of a precursor into a heated surfactant solution as part of the approach was also highlighted by Tran *et al.* Small, uniformly sized NPs with limited size distributions were rapidly generated due to this process. They are also known as homogenous NPs because of their homogeneity. Thermal breakdown happens at different temperatures and pressures depending on the nature of the metal ions and the ligands in coordination compounds.

Coordination chemicals can be both thermodynamically and kinetically inert. The thermal breakdown does not necessitate the use of a stabilizer. Carboxylic acids and alkylamines have been shown to influence the production of monodispersed NPs derived *via* thermal breakdown by Rao and colleagues. The influence on nanoparticle synthesis results from the total effect [15].

Ultrasonication

Another approach for preparing nano emulsions with high control over emulsion properties is ultrasonication. It is possible to use it to create a nano emulsion in real-time, as well as to reduce the size of an emulsion that has already been produced. Cavitation happens when ultrasonic waves flow through an emulsion, causing microbubbles/cavities to form, expand, and collapse. Cavitation happens when ultrasonic waves flow through an emulsion, causing microbubbles/cavities to form, expand, and then collapse. An emulsion is cavitated when ultrasonic waves pass through it. This causes microbubbles/cavities to sprout, expand, and explode. A localized hotspot with temperatures up to 5000K and pressures up to 1000 bar results from these transitory collapse settings. Emulsion processes might benefit from cavitation conditions of this intensity. There are two methods for ultrasound-based emulsification. The acoustic field is where the first generation of droplets is generated. Asymmetric cavity collapse causes high turbulence and microjets, which break up and disperse droplets during the continuous phase. Numerous studies have shown ultrasound to produce nano emulsions with droplet sizes as small as 100 nanometers. Emulsions with smaller droplet sizes are more stable over time. As a consequence, ultrasounds are better able to manage particle size distribution and increase emulsion stability. An additional popular approach for emulsification is high-pressure homogenization in industries like pharmaceutical, food, and biotechnology.

High-pressure homogenizers combine the oil, surfactant, and water in a shear flow field with high and high turbulence. There are droplets smaller than 100 nm in the dispersed phase that is broken up by the turbulent flow. The dynamic equilibrium between breaking and coalescence is controlled by the relative velocities of the droplets, resulting in homogeneous droplets with increased shelf life and texture [16].

Physical Methods for the Synthesis of NPs

Laser Ablation

Laser ablation (LA), a time-consuming procedure, removes tissue. Depending on the wavelength and refractive index of the target material, the laser penetrates the sample surface to varying degrees. To extract electrons from bulk material, laser light creates a strong electric field. Free electrons are generated, and energy is transferred when they come into contact with other atoms in the bulk sample *via* collision. Thus, vaporization occurs on the surface as a consequence. All kinds of material—atoms and molecules in particular—will transition from their normal state to a plasma state when the laser flux is strong enough. The seed plasma expands and cools rapidly because of the pressure difference between it and the atmosphere. In a vacuum or a gaseous atmosphere, LA can take place. Pulsed LA is a combination of LA with a tube furnace. This approach allows for more exact temperature, gas type rate, and pressure control during growth [17].

Mechanical Milling

Mechanical milling is the most basic top-down manufacturing approach, whether or not a chemical reaction occurs in the solid state of materials. The milling method (ball or attrition miller), power employed, milling medium (*e.g.,* tungsten carbide ball), process control solution (*e.g.,* toluene), speed (revolutions per minute), and duration all affect ENM product properties in basic mechanical milling.

Metal, oxide, or complex metal NPs are produced in mechanochemical milling once the milling process has been completed. It is used in this process to create the desired nanomaterial composition by combining metals, alloys, and other powder mixtures with appropriate reactants to aid or complete the solid-state reaction, along with surface modifying agents (*e.g.,* carboxylic acid or other acids) and process control solutions such as stearic acid or toluene to achieve the desired results. Rather than being used for specific/precision applications, these end products are more commonly used for bulk nano-grained materials or nanocomposites because of the broad size dispersion and varied shape. To produce NPs or nanowires, solid-state processes can be carried out in joule furnaces at high temperatures (*e.g.,* 850 °C) without milling. Nonylphenol ethers and other nanomaterials need a precursor material, solutions that help form these compounds, and a process control solution. There must be a thorough blending of the precursor materials to ensure that the final result is homogeneous [18].

Sputtering

In order to vaporize an object, an object must be bombarded by atom-sized particles that are powerful enough to vaporize the object's atoms physically (rather than thermally). This particle is often an ionized gaseous material charged in an electric field. Sputtering and sputter deposition have a long and illustrious history [19]. Sputtering is a thin-film manufacturing technique in various industries, including semiconductor processing, surface finishing, and jewelry fabrication. Metal deposition is the most common industrial application, but it has also been utilized to make insulating materials. To remove atoms from a surface, an ionized atom must be accelerated into the surface. A thin film of the ejected material may be created by condensing the atoms onto a sample. This process is referred to as sputter deposition (sputtering).

The same physical procedure can eliminate an undesired material from a sample. The expelled atoms might be gathered on the shielding of the chamber. Sputter etching is the name for the latter method [20].

Electron Beam Evaporation

Electron beam evaporation evaporates the source material by directing a strong beam of electrons with high energy. The thermionic emission of electrons from a heated filament may be used to speed and accelerate the evaporation of any substance. Typically, 1 A of emission is accelerated by a voltage drop of 10 kV to generate 10 kW. In an electric magnetic field (E), the Lorentz force (F) on an electron is given by:

$$F = F_E + F_B = q_e E + q_e (V \times B) \tag{3}$$

Where F is in N, qe in C, E in V/m, B in webers/m^2 5 teslas, and the electron velocity v is in m/s. The cross-product vector, FB, is perpendicular to both v and B. electrons are accelerated away from the filament or cathode according to equation (3). According to the second force term, when electrons cross the magnetic field lines at this speed, they are deflected sideways. For the second force to be equal, electrons' centrifugal force must be equal [21].

Electro Spraying

In certain ways, electrospray and electrospinning are both forms of the electrohydrodynamic process. A high-voltage power supply, a syringe filled with a precursor solution (typically polymer solution) and metallic needle, a syringe pump regulating the solution feeding rate, and a ground collector are utilized. Electrospray generates Taylor cones stabilized by liquid surface tension, electrostatic forces, and gravitational pulls. Spherically shaped jetting beads are created by an electrostatic force acting as a counterbalance to the surface tension of emitting droplets.

Particulate rather than fibrous products may be produced using electrospray rather than electrospinning because of a lower degree of electrostatic stretch. Electrospray requires a certain concentration (and viscosity) of polymer solution to establish the geometry of the electrospray products. As a result, the polymer concentration in the electrospray polymer solution is low. Once they reach the collection, droplets of a low-viscosity solution are still liquid. Intermittent fiber production will result from the solution's high viscosity.

Simple electrospray and coaxial electrospray are two types of electrosprays that use different spinnerets. Coaxial electrospray is more difficult since it involves two incompatible liquids. When two immiscible liquids mix in the spinneret, electrostatic forces and solution surface tension generate a conically formed cone-jet. To account for the different tangential electrostatic forces exerted by each immiscible liquid, the fastest electrical relaxation liquid (typically water phase) is the principal driving force behind deforming compound jets during coaxial electrospray. This liquid is referred to as the primary driving phase. It is possible to create core-shell structures using coaxial electrospray with the driving liquid within. Core-shell particles with a definite structure may be generated in other ways except *via* coaxial electrospray. Electrospray with a water-in-oil emulsion produces microparticles with numerous cores, as well. Because of their simplicity and adaptability, microparticles with adjustable architectures may be easily and quickly created [22].

Electrochemical Methods for the Synthesis of NPS

Electrochemical methods have generated various metallic, semiconducting, and oxide NPs. In the future, electrochemical synthesis has the potential to be inexpensive and high-yielding. Furthermore, electrochemical synthesis techniques allow for both batch and continuous operations.

Fig. (2) shows a diagram of the electrochemical production of metallic NPs stabilized by tetraalkylammonium ligands. Each electrode's reactions are listed below:

Anode: $M_{bulk} \rightarrow M^{n+} + ne^-$

Cathode: $M^{n+} + ne^- + stabilizer \rightarrow M_{np}$ stabilizer

Where M_{bulk} and M_{np} refer to bulk and NP metals, respectively. The tetraalkylammonium salt ($R4N+X^-$), where X =Cl, Br, or I and R = n-C$_m$H$_{2m+1}$), is used as an NP stabilizer. Many factors influence the electrochemical synthesis process' particle size selectivity, including the polarity of the solvent, the density of the current, the charge flow, the distance between electrodes, and the temperature.

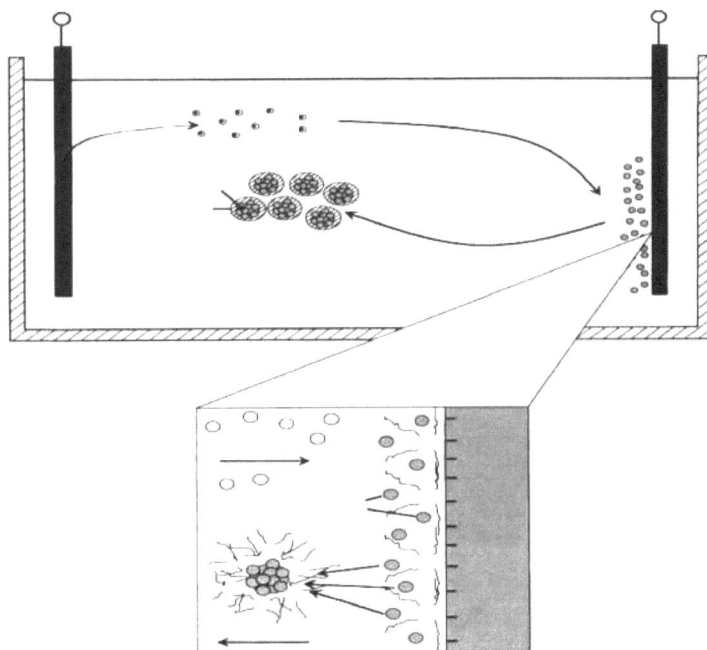

Fig. (2). Schematic design of an electrochemical setup used to synthesize $R_4N^+X^-$ stabilized transition metal colloids [23].

Temperature, charge flow, and current density variable factors are employed to tune NP size from one to five nanometers in the case of Pd. To create bimetallic NPs, such as Ni–Pd, Fe–Co, and Fe–Ni, the method has also been utilized to synthesize NPs. Cobalt NPs were electrochemically synthesized at 2 to 7 nm with adjustable average particle size. An average current density of roughly three mA/cm^2 yields the tiniest particles. As the current density decreases, the average particle size increases. The yield and scalability of this method remain undetermined. Electrochemical methods are also used to deposit these NPs into the substrate, resulting in a variable film. A low-cost pulsed electrodeposition technique is used to manufacture various metallic NPs of varied sizes. It can modify the NPs' physical characteristics by altering the pulse shape, organic additives, bath temperature, or pH [24].

Inert Gas Condensation

Nanostructured materials may be synthesized bottom-up using inert gas condensation (IGC). Initially, the material is vaporized, then swiftly condensed to get the appropriate particle size.

At a certain temperature, evaporation takes place in a crucible heated by a graphite heater. An oil diffusion pump in this device evacuates the chamber to a pressure of around 2×10^{-6} Torr. Typically, the pressure range is between 0.01–0.4 Torr, and the crucible is heated fast. There are low-pressure leaks of inert gas entering the chamber after evacuation (usually of He, Xe, or Ar). A water-cooled surface is used to create, cool, and collect ultra-fine metal particles.

Carbon-coated electron microscope grid connected to a water-cooled surface in the center may collect powder particles and be examined in a transmission electron microscope. It was feasible to sample particles on the grid less than a monolayer thick with a shutter.

The collected nanophase particles are condensed in a compactor powder pellet available in diameters ranging from 8–9 mm and a thickness ranging from 0.1–0.5 mm, depending on the manufacturer. Compaction equipment is not necessary if you are just interested in nanostructured powders. It is a common practice to cool down cluster collecting devices with liquid nitrogen to enhance heat transfer and produce a significant temperature differential. A low-pressure, high-purity inert gas may be back-filled to a vacuum of less than 10^{-5} Pa using a turbo-molecular pump. The metal is evaporated in the process.

During the collision of metal atoms with inert gas molecules, loose powder crystals are created. This process results in particle production, nucleation, and high supersaturation levels. There are convection currents that move fine powder from the crucible region to the collecting device due to the evaporation source (cold finger) heating and cooling inert gas [27].

Vapor Deposition

The bottom-up method known as chemical vapor deposition (CVD) might produce graphene nanostructures. In order to manufacture materials based on graphene, research has been conducted on CVD and other surface precipitation techniques. Substrates are typically exposed to a volatile precursor in a CVD process of this type. When the precursor reacts or decomposes, it leaves behind the desired product, then applied to the substrate. In order to synthesize graphene using the CVD method, the first step is to pyrolyze the precursor material to obtain carbon, which is then used to make graphene. It is feasible to minimize the creation of carbon soot in the gas phase by using metal catalysts. Although this process requires high temperatures, it is still within one's control. When working with substrates based on transition metals, an acidic solution may dissolve the newly created graphene, which can then be transferred to a different substrate.

Product purity and fine structural regularity are two of the CVD method's advantages. Newer CVD methods have been developed to deal with extreme heat. For a uniform product coating, CVD procedures like ultra-high-vacuum CVD or low-pressure CVD decrease the requirements and eliminate undesirable reactions [28].

Arc Discharge

The anode and cathode electrodes, made of high-purity graphite, are kept at short distances in a helium atmosphere while conducting an arc discharge. A hard cylindrical deposit is formed on the cathodic rod due to carbon evaporating from the anode and re-condensed under these circumstances. The arc-evaporation process relies heavily on the amount of current being used. The greater current application results in a hard, sintered material with few free nanotubes. Consequently, it is recommended that the current be minimized. Arc discharge can create nanotubes up to a few hundred microns in length. When scaling up the arc discharge process, a limited number of metal catalysts are required, which boosts nanotube yields. The final product also has amorphous carbons, non-tubular fullerenes, and catalyst particles. Because of this, extra purifying methods are needed. High temperatures are also required for this procedure. Arc discharge requires a temperature of 600-1000°C. Therefore, the tubes may have different lattice layouts. Nanotube chirality and diameter may be challenging to regulate, as well [29].

Green Methods for Synthesis of NPs

Bacteria

In commercial biotechnological applications, including bioremediation and genetic engineering, bacteria species have been used. In nanoparticle synthesis, bacteria are excellent choices because they can decrease metal ions. Various bacterial species are used to prepare metallic and other new NPs. Prokaryotic bacteria and actinomycetes have frequently produced metal/metal oxide NPs. The simplicity with which bacteria may be manipulated has led to the widespread use of bacterial nanoparticle production. *Escherichia coli, Lactobacillus casei*, and *Aeromonas sp.* are examples of microorganisms. They have been widely employed to manufacture bio-reduced silver NPs with various size/form properties. *SH10 Antarctica, Phaeocystis, Enterobacteriaceae, Geobacteriaceae, Arthrobacteriaceae, Corynebacteriaceae SH09*, and *Shewanella oneidensis* are all examples of microbial species that may be found in the guts of humans and other animals. They all have a role in the digestion of food and the removal of waste. *E. coli, Bacillus subtilis*, and *Shewanella algae*, in addition to *Desulfovibrio desulfuricans*, have all been employed extensively in producing gold NPs. There have also been experiments with *Rhodopseudomonas* capsulate and *Plectonema boryanum* [30].

Fungi

Monodispersed metal/metal oxide nanoparticles may be produced through a fungi-mediated synthesis of metal/metal oxides. Their various intracellular enzymes make them superior biological agents for producing metal and metal oxide NPs. Competent fungi than bacteria more readily produce NPs. Enzymes and proteins/reducing components on the surface of fungi provide them various advantages over other species.

If metallic NPs may be formed within the fungal cell wall or inside the fungal cell itself, the most likely mechanism is enzymatic reduction (reductase). It is possible that a wide variety of fungi make use of metal and metal oxide NPs (including silver and gold) [30].

Yeast

Eukaryotic cells include single-celled microorganisms called yeasts. Different types of yeasts may be found in various sizes, shapes, and colors. Different species are used to make a huge number of metallic NPs. Many research groups have employed yeast to manufacture NPs and nanomaterials effectively. Biosynthesis of silver and gold NPs was carried out using a Saccharomyces cerevisiae broth and a silver-tolerant yeast strain [30].

Plants

Heavy metals can accumulate in numerous parts of a plant's body. Due to the ease and cost-effectiveness of plant extract, biosynthesis procedures have been deemed a good alternative to conventional nanoparticle production methods. In the "one-pot" synthesis technique, metallic NPs can be reduced and stabilized using various plants. Plant leaf extracts are being used more and more in metal/metal oxide nanoparticle production studies as a green synthesis approach. Metal salt can be reduced to NPs by plant biomolecules (such as carbohydrates, proteins, and coenzymes). Extracts from plants were the first to investigate gold and silver metal NPs, just like other biosynthetic processes. Various plants (including *aloe vera, oat, alfalfa, Tulsi, Lemon, Neem, Coriander, Mustard,* and *lemongrass*) have been utilized to synthesize silver and gold NPs. There are ways to make metal nanoparticles from soluble salts that have been absorbed by plants in living organisms. In this type of research, NPs have been synthesized *ex vivo*.

As well as *in vivo* NPs of zinc, nickel, copper, and cobalt were found in *alfalfa* and *sunflower*. Plant leaf extracts such as *coriander, crown flower, copper leaf, China rose, Green Tea,* and *aloe leaf* broth extract have also been used to make ZnO NPs. Iravani's study provides a detailed look at the plant materials that may be utilized to create NPs [30].

BIOLOGICAL SYNTHESIS OF NPS FROM PLANTS AND MICROORGANISMS

NPs Synthesis using Microorganisms

Gold NPs

Even in ancient Rome, when they were first used to decorate glassware, gold NPs (AuNPs) had a long and distinguished history in the chemical sciences. For centuries, AuNPs have been used to cure many diseases. It was Michael Faraday, about 150 years ago, who launched contemporary AuNPs production by recognizing that colloidal gold solutions exhibit characteristics distinct from bulk gold. In light of the growing need for environmentally friendly materials, bio nanotechnology has gained much interest in nanotechnology. Gold NPs have been created *ex vivo* by the fungus *Fusarium oxysporum* and the actinomycete *Thermomonospora sp.* In addition, researchers discovered that the fungus *Verticillium* sp. might produce gold NPs inside its cells. By treating bacterial cells with Au^{3+} ions, Southam and Beveridge have demonstrated that gold nanoparticles may be precipitated inside cells. The alkalotolerant *Rhodococcus* sp. was used to manufacture monodisperse gold NPs under alkaline and slightly increased temperature conditions. *Filamentous cyanobacteria* have been shown to produce gold nanostructures using Au(I)-thiosulfate and Au (3)-chloride complexes. Nair and Pradeep used *Lactobacillus* to generate nanocrystals and nanoalloys.

Silver NPs

Silver nanoparticle antibacterial activity is effective against Gram-positive and Gram-negative bacteria, including highly resistant strains like *methicillin*-resistant *Staphylococcus aureus*. Biomimicry approaches for developing sophisticated nanomaterials have been developed due to studying nature's secrets. Tiny silver particles may be manufactured utilizing microorganisms in ecologically friendly nano factories. Silver NPs, the majority of which are spherical, are produced by bacteria decreasing Ag+ ions. An isolated silver mine strain of bacteria known as *Pseudomonas stutzeri AG259* has been found to play an important role in the reduction of Ag+ and the formation of well-defined silver NPs in the periplasmic space of the bacteria when placed in a concentrated aqueous solution of silver nitrate. *Verticillium spores, Fusarium oxysporum*, or *Aspergillus flavus* were used as a film, solution, or cell surface accumulation to generate AgNPs.

Alloy NPs

Nanoparticle alloys, used in various industries, are of particular interest. Senapati *et al.* claim that the bimetallic Au-Ag alloy was created using *F. oxysporum*. A secreted cofactor called NADH is critical to the composition of Au-Ag alloy NPs, they claim. Zheng *et al.* investigated the creation of Au-Ag alloy NPs using yeast cells. An *ex vivo* approach was used to create the bulk of the Au-Ag alloy NPs. As characterized by fluorescence microscopy and electron microscopy, they were found to be mainly polygonal NPs. With the vanillin sensor, the electrochemical response was five times greater than with ordinary glassy carbon electrodes. The *Fusarium semitectum* was used by Sawle *et al.* to make core-shell Au-Ag alloy nanoparticles. An NPs suspension prepared this way might be maintained stable for many weeks.

Other Metallic NPs

It is well-known that microorganisms can damage heavy metals. Anti-heavy metal resistance in bacteria is often achieved by combining two strategies: chemical detoxification and energetic ion efflux from the cell using membrane proteins functioning as ATPase, chemical cation, or proton anti transporters. The solubility of microbial resistance is also affected by changes. The metal ion-reducing bacteria *Shewanella algae* were utilized to manufacture platinum NPs by Konishi and his colleagues. *S. algae* resting cells transformed aqueous $PtCl_6^2$ ions to elemental platinum in 60 minutes when lactate was utilized as an electron donor at ambient temperature and neutral pH. There were platinum NPs in the periplasm with a diameter of around 5 nm. *Enterobacter sp.* cells were capable of producing mercury NPs, as reported by Sinha and Khare. To create mercury NPs with a diameter of 2–5 nm and mono dispersion, growth parameters (pH eight and lower mercury content) must be satisfied. U(VI), Tc (VII), Cr (VI), Co (III), and Mn were all found to be reduced by *Pyrobaculum islandicum* and anaerobic *hyperthermophilic* bacterium when hydrogen was used as an electron donor (IV). It is possible to generate palladium NPs using *Desulfovibrio desulfuricans* (a sulfur-reducing bacterium) and *S. oneidensis* (a metal ion-reducing bacterium).

Oxide NPs

Oxide NPs are a common form of compound nanoparticle that microorganisms create. Magnetic and nonmagnetic oxide NPs produced in biological systems were the focus of this section. Most magnetotactic bacteria examples are utilized to create magnetic oxide NPs. In contrast, biological systems are used to create nonmagnetic oxide NPs [31].

Metal NPs Synthesis using Plants

Plants are preferred over bacteria for green NPs manufacturing because they are nonpathogenic, and several methods have been thoroughly investigated (Fig. **3**). A wide variety of plants have been employed to make metal NPs. Compared to their bulk counterparts, these NPs exhibit unique optical, thermal, magnetic, physical, chemical, and electrical properties. This technology has no trouble with industry-specific applications. AgNPs are made from a variety of biological organisms. In 4 hours, *Jatroa curcas* extract generates homogeneous (10–20 nm) AgNPs from $AgNO_3$ salt. Leaf extracts from *Acalypha Indica* may be used to make AgNPs. The AgNPs had a size range of 20 to 30 nm and were homogeneous. In another work, AgNPs were produced from *Medicago sativa* seed exudates. As NPs were detected within a minute of metal salt contact and 90% of Ag^+ was decreased in production at 30°C, the decline in Ag^+ happened quickly. The particles were stabilized by a leaf broth component, making them spherical. Fast-growing Ag NPs were created by using the fruit extract of *Terminalia chebula*. Extracts from *Eucalyptus macrocarpa* leaf and *Nyctanthes arbore tristis* flower generated Ag and Au NPs with cubic shapes ranging in size from 50 to 200nm, respectively, whereas the *coriander* leaf extracts produced Ag and Au NPs with sizes of 7–58nm. *Phyllanthin*, a pigment produced by the *Phyllanthus amarus* plant, may be used to make gold and silver NPs. Unlike prior experiments, this one does not use the plant as a whole to manufacture metallic NPs. Metallic NPs were synthesized using just one component of a plant extract.

The form and size of the NPs were modified by the amount of *phyllanthin* utilized. Low *phyllanthin* concentrations produced AuNPs with triangular and hexagonal shapes, but larger concentrations produced NPs with a more spherical shape. Plant-derived polysaccharides and phytochemicals, soluble starch, cellulose, dextran, chitosan, alginic acid, and hyaluronic acid may all be used to produce silver and gold NPs effectively. These compounds have the advantage of using less harmful chemical compounds while also allowing for creation of nanocomposites with various metals. Spherical AuNPs aggregated and produced a new type of nano triangles with the aldehydes and ketones in the *lemongrass* plant extract after combining it with gold tetrachloride solution. When silver gold salts are added to *Azadirachta indica* leaf broth, a complex emerges.

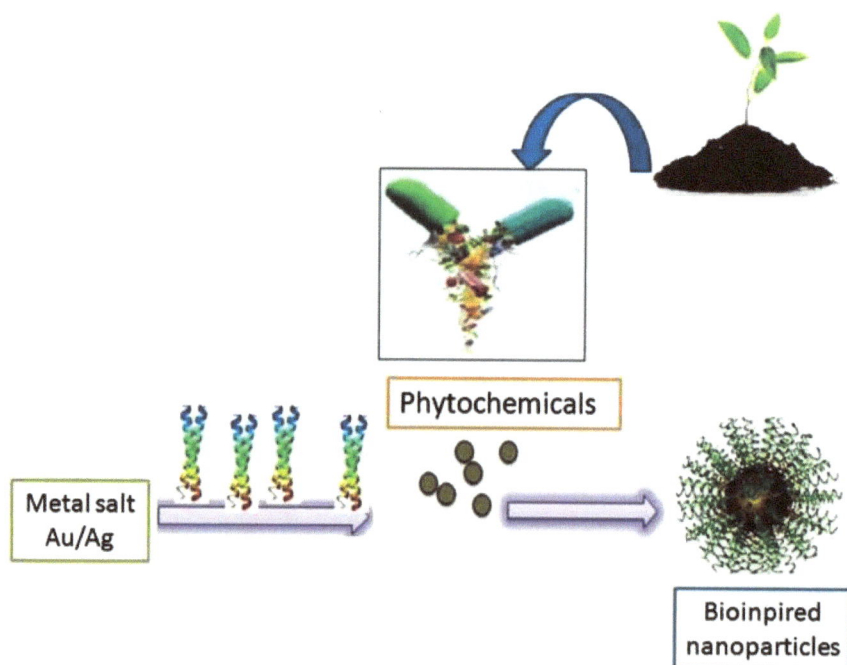

Fig. (3). Green synthesis of Cds NPs by bacteria [32].

It is, therefore, possible to produce silver, gold, and bimetallic silver-gold NPs concurrently by combining the two metallic ions. Within two hours, the pace of nanoparticle production had reached a plateau. The leaf's terpenoid and flavanone components were credited for the NPs' long-term stability.

Antibiotic-resistant *Staphylococcus aureus* may benefit from the increased bactericidal activity of phytochemically reduced NiO NPs with *garlic* and *ginger*. Bimetallic NPs were made by alloying silver and gold. A plant extract and one metallic ion precursor are reduced in competition to make these compounds. The core-shell structure of the Ag-Au NPs is made of Au because of its greater reduction potential. As the Ag ions are reduced, they form a shell, and the core is solidified. *Azadirachta indica, Anacardium occidentale, Swieteni amahagony*, and *cruciferous* vegetable extracts have all been successfully employed to manufacture Ag-Au bimetallic NPs.

Copper (Cu) NPs and copper oxide were created using extracts from several plants (CuO). *Magnolia Kobus* leaf extract and *Syzygium aromaticum* (Clove) were used to make Cu NPs ranging from 40 to 100 nm, with a spherical to granular form and an average particle size of 40 nm. *Euphorbia nivulia* (Common *milk hedge*) *Latex* and *Sterculia urens* (*Karaya gum*) Latex stems were utilized to manufacture CuO NPs stabilized and coated with terpenoids and polypeptides from the Latex, respectively, in this work. The resulting NPs, which were particularly stable, had an average particle size of just over 4.8 nanometers. Song *et al.* helped demonstrate the first platinum nanoparticle synthesis. Carboxylic acids, amines, and alcohols from *Diospyros kaki* (*Persimmon*) leaf extract. Pt ions are reduced by the leaf extract's ketone functional group. In less than 2.5 hours, 90% of Pt ions were reduced to NPs. In order to rule out the possibility of an enzyme-mediated process, the experiment was conducted at 95°C.

The *cinnamon zeylanicum bark* and *Annona squamosa* (*custard apple*) extract were utilized to make Pd NPs with a diameter of 75–85nm, according to the research.

Soy leaf extract was used to make NPs with an average diameter of 15 nm (Glycine max). To synthesize face-centered cubic crystal palladium NPs, extracts from commercially available tea and coffee plants, *Camellia sinensis* and *Coffea arabica*, were used (Coffee). *Gardenia jasminoides* (*Cape jasmine*) extract contains antioxidants (*geniposide, chlorogenic acid, crocins, and crocetin*) that stabilize and reduce the synthesis of palladium NPs. Other plants, such as *Ocimum sanctum* leaf extract (*Holy basil*), plant wood nanomaterial, and *red pine lignin*, produced platinum and palladium NPs (*Pinus resinosa*).

Many plant extracts, including *Annona squamosa peel, Cocos nucifera coir*, and the leaves from *Nyctanthes arbor-tristis Psidium Gujava* and *Catharanthus roseus*, were used to produce tiny spheres of titanium dioxide (TiO_2) effectively. *Physalisalke kengi* and *Sedum alfredii* latex, *Aloe vera, Calotropis procera*, and *Calotropis Procera* latex were all employed to make zinc oxide NPs. Indium oxide spherical NPs with diameters ranging from 5 to 50 nm were created using *aloe vera* leaf extracts (*Aloe barbadensis*). *Sorghum bicolor bran* extracts and leaves from *Euphorbia milii, Tinospora cordifolia*, and *Datura innoxia* extracted Fe NPs using green chemistry techniques. With sizes between 10 and 12.5 nm, the Latex of the *Jatropha curcas* plant produced Pb NPs. Metallic NPs can be synthesized from plant extracts or whole plant extracts. Living plants can also produce metallic NPs. By cultivating *Arabidopsis thaliana* in its regular medium, a new method for producing PdNPs has been discovered. For the next 24 hours, the salt solution was replaced with potassium tetrachloropalladate (K_2PdCl_4). Transmission electron microscopy revealed PdNPs with a 2–4 nm size range. In Suzuki-Miyaura coupling processes, biologically produced PdNPs outperformed commercially available PdNPs.

In order to produce AuNPs, *Alfalfa* seeds were cultivated for two weeks in varying concentrations of $K(AuCl_4)$. Commercial viability is limited by the two-week time required to produce NPs using this technology.

Even if the production time is shortened, it is still possible. This might be a great way to produce NPs cheaply and with no environmental impact [32].

FACTORS AFFECTING BIOLOGICAL SYNTHESIS OF METAL NPS

Influence of pH

In green technology techniques, the pH of a solution influences the production of NPs. For the first time, NPs size and texture can be influenced by their solution medium's pH, according to new research. The pH of the solution medium may be used to manipulate NPs size. Silver NPs produced by Soni and Prakash were studied for their form and size in relation to pH [33].

Influence of Reactant Concentration

Nanoparticle production may be affected by the number of biomolecules in plant extracts. According to Huang *et al.*, increased concentrations of *Cinnamomum camphora* (*camphor*) leaf extract in the reaction fluid considerably impacted the formation of Au and Ag NPs. The NP's form changed from triangular to spherical when chloroauric acid's extract content was increased. Changing the chloroaurate-ion ratio of gold triangular plates to spherical NPs in *Aloe vera* leaf extract, Chandran *et al.* may be able to do this. The study found that the extract's carbonyl components influenced particle formation as well. Extract concentration affected particle size. Particles ranging from 50 to 350 nm were observed. Changing the concentration of *Plectranthus amboinicus* leaf extract produced Ag NPs with decahedral, hexagonal, triangular, and sphere morphologies [34].

Influence of Reaction Time

The reaction medium's incubation period considerably affects the quality and kind of NPs manufactured utilizing green technology. As with produced NPs, their qualities evolved with time.

The synthesis process, light exposure, and storage conditions significantly impacted the results. Particles can clump together owing to long-term storage; the particles may shrink or expand; they may have a shelf life, and so on, all of which impact their usefulness [33].

Influence of Reaction Temperature

Each of the three methods relies heavily on temperature to produce NPs. Regarding procedures, the physical ones necessitate the highest temperatures (over 350°C), whereas chemical ones necessitate temperatures below 350°C. To synthesize NPs with green technology, you will often need temperatures below 100 degrees celsius. The type of nanoparticle generated is dependent on the reaction medium's temperature [33].

Antibacterial Activity of NMNPs Synthesized using Plants Extract

As a natural and inherent phenomenon, resistance to antimicrobials can be acquired or transferred. For example, bacterial species may be able to withstand or diminish the efficacy of antibiotics because of their inherent functional or structural properties. The three phases of the development of drug resistance are the acquisition, expression, and selection of bacteria with resistance genes. As a first step, bacterial drug resistance is acquired by horizontal gene transfer (transduction, transformation, and conjugation) (HGT). Antibacterial agents - lactams and fluoroquinolones are under threat from HGT. If a random mutation occurs in an already-existing genetic component, bacteria may acquire a new, resistant gene. For example, antibiotic-resistant bacteria might acquire resistance to a new antibiotic. This is known as multiple drug resistance (MDR). Second, bacteria exposed to antimicrobial agents produce the resistance gene. Thirdly, resistance to antibiotics is frequent when the microorganisms that express resistance genes can proliferate in a suitable environment. Drug resistance may be encouraged by an antibiotic with a long half-life and poor patient compliance.

If bacteria are exposed to antibiotics without eradication, they are subjected to conditional or selective pressure. Using antibiotics for longer durations increases the risk of developing resistance. It is possible that some bacteria can re-grow and become resistant to bacteriostatic drugs, which do not kill but inhibit germs instead. As a result of patient noncompliance, resistance genes can evolve and/or acquire. Antimicrobial resistance in bacteria is a multifaceted process.

P. aeruginosa and *E. coli's* low antibiotic susceptibility is due to their drug efflux mechanism. An increase in the rate at which drugs exit the bacterial cell is the primary cause of this effect. The transmembrane efflux pump prevents the antimicrobial agent from reaching the bacterial cell's hazardous concentration level. There is no difference in periplasmic space between Gram-negative and Gram-positive bacteria regarding volume. An inner membrane is surrounded by a *peptidoglycan* cell wall.

A linker protein in *P. aeruginosa's* drug efflux pump connects an H+/drug antiporter protein in the periplasmic region to an outer membrane channel protein. These efflux proteins were overexpressed in *P. aeruginosa*. Genes encoding efflux proteins are most often affected when the regulatory protein needed to suppress them is mutated. Similarly, *E. coli* utilizes the drug efflux system. The transmembrane proton gradient powers at least nine pumps, each expelling a different kind of antibiotic. As a result, *E. coli* is now resistant to various drugs. In Gram -ve bacteria, a lipopolysaccharide-based outer membrane is widespread. A peptidoglycan-based cell wall covers just one plasma membrane in Gram +ve bacteria. Because Gram-negative bacteria are more tolerant to antibiotics than Gram-positive bacteria are, this helps explain why this is the case. *MRSA* and *Klebsiella pneumoniae* are examples of bacteria that have developed resistance to methicillin and other antibiotics by enzymatic inactivation of antibiotics, covalent modifications to antibiotics, mutations to antibiotic targets, and the protection of antibiotic targets.

According to a few studies, bacteriophage MNP resistance can be traced back to the production of extracellular molecules that lead to MNP aggregation and precipitation. In the last several weeks, three *E. coli* have been treated with AgNPs at sub-inhibitory levels. Antibiotic resistance was generated as a consequence of several interactions with AgNPs. Due to the prolonged exposure, the bacteria developed resistance to the AgNPs' effects. This was achieved as a result of the bacteria producing flagellin. This adhesive protein causes AgNPs to precipitate by weakening their stability. As a result, the antibacterial activity of AgNPs is lost since they cannot enter the bacteria cell. Another way bacteria become resistant to MNPs is by forming biomolecule coronas. It is common in bodily locations like the intestines, lungs, and open sores. In order to prevent the nano antibiotic from binding to harmful bacteria, a biomolecule corona is produced.

The bacterium's capacity to alter its surface charge as a defensive tactic makes it resistant to MNPs' bactericidal effects. A change in the phospholipid structure modifies the bacteria's surface charge, which is how this is performed. To counteract the growing problem of antibiotic resistance, scientists are now looking into the potential of plant-mediated antimicrobials [36].

CONCLUSION

Nanomaterials, in general, and noble metal NPs, in particular, are interesting materials for many applications. In comparison with other nanoscale systems, noble metal NPs are much more stable and exhibit much less cytotoxicity. Hence, the applicability of these NPs is far greater than that of other nano systems, especially in biology. Increasing awareness of green chemistry and biological processes has led to a desire to develop an environment-friendly approach to synthesizing non-toxic NPs. Unlike other processes in physical and chemical methods, which involve hazardous chemicals, microbial biosynthesis of NPs is a cost-effective and eco-friendly approach. Therefore, microbial synthesis of NPs has emerged as an important branch of nanobiotechnology. Due to their rich diversity, microbes have an innate potential for the synthesis of NPs, and they could be regarded as potential bio factories for nanoparticle synthesis. However, to improve the rate of synthesis and monodispersity of NPs, factors such as microbial cultivation methods and downstream processing techniques have to be improved, and the combinatorial approach, such as photobiological methods, may be used. The delineation of specific genes and characterization of enzymes that are involved in the biosynthesis of NPs is also required. Thus, complete knowledge of the underlying molecular mechanisms that mediate the microbial synthesis of NPs is mandated to control the size and shape as well as crystallinity of NPs. Future research on microbes-mediated biological synthesis of NPs with unique optoelectronics, physicochemical and electronic properties is of great importance for applications in chemistry, electronics, medicine and agriculture.

REFERENCES

[1]   T.K. Sau, A.L. Rogach, F. Jäckel, T.A. Klar, and J. Feldmann, "Properties and applications of colloidal nonspherical noble metal nanoparticles", *Adv. Mater.*, vol. 22, no. 16, pp. 1805-1825, 2010.
[http://dx.doi.org/10.1002/adma.200902557] [PMID: 20512954]

[2]   G. Vinci, and M. Rapa, "Noble metal nanoparticles applications: Recent trends in food control", *Bioengineering (Basel)*, vol. 6, no. 1, p. 10, 2019.
[http://dx.doi.org/10.3390/bioengineering6010010] [PMID: 30669604]

[3]   A. Martínez-Abad, "Silver-based antimicrobial polymers for food packaging", *Multifunct. Nanoreinforced Polym. Food Packag*, pp. 347-367, 2011.
[http://dx.doi.org/10.1533/9780857092786.3.347]

[4]   S. Rahim, F. Jan Iftikhar, and M.I. Malik, "Chapter 16 - Biomedical applications of magnetic nanoparticles", In: *Metal Nanoparticles for Drug Delivery and Diagnostic Applications* Elsevier Inc., 2019, pp. 301-328.

[http://dx.doi.org/10.1016/B978-0-12-816960-5.00016-1]

[5]   Ita Kevin, "Microemulsions", In: *In: Transdermal Drug Delivery*, 2020, pp. 97-122.
        [http://dx.doi.org/10.1016/B978-0-12-822550-9.00006-5]

[6]   M.A. Malik, M.Y. Wani, and M.A. Hashim, "Microemulsion method: A novel route to synthesize
        organic and inorganic nanomaterials", *Arab. J. Chem.*, vol. 5, no. 4, pp. 397-417, 2012.
        [http://dx.doi.org/10.1016/j.arabjc.2010.09.027]

[7]   E. Suvaci, and E. Özel, "Hydrothermal Synthesis", *Encycl. Mater. Tech. Ceram. Glas*, pp. 59-68,
        2021.
        [http://dx.doi.org/10.1016/B978-0-12-803581-8.12096-X]

[8]   G. Gadea, A. Morata, and A. Tarancon, *Semiconductor Nanowires for Thermoelectric Generation.* vol.
        98. 1st ed. Elsevier Inc., 2018.
        [http://dx.doi.org/10.1016/bs.semsem.2018.01.001]

[9]   V. Sáez, and T. Mason, "Sonoelectrochemical synthesis of nanoparticles", *Molecules,* vol. 14, no. 10,
        pp. 4284-4299, 2009.
        [http://dx.doi.org/10.3390/molecules14104284] [PMID: 19924064]

[10]   A. Morsali, and L. Hashemi, *Nanoscale coordination polymers: Preparation, function and
        application.* vol. 76. 1st ed. Elsevier Inc., 2020.
        [http://dx.doi.org/10.1016/bs.adioch.2020.03.007]

[11]   T.V. Gavrilović, D.J. Jovanović, and M.D. Dramićanin, *Synthesis of multifunctional inorganic
        materials: From micrometer to nanometer dimensions.* Nanomater. Green Energy, 2018, pp. 55-81.
        [http://dx.doi.org/10.1016/B978-0-12-813731-4.00002-3]

[12]   D. Sumanth Kumar, B. Jai Kumar, and H.M. Mahesh, "Chapter 3 - Quantum nanostructures (qds): an
        overview", In: *Advances and key technologies micro and nano technologies* WP, 2018, pp. 59-88.
        [http://dx.doi.org/10.1016/B978-0-08-101975-7.00003-8]

[13]   W. I. Wu, P. Rezai, H. H. Hsu, and P. R. Selvaganapathy, "Materials and methods for the
        microfabrication of microfluidic biomedical devices", *Microfluidic Devices for Biomedical
        Applications,* pp. 3-62, 2013.
        [http://dx.doi.org/10.1533/9780857097040.1.3]

[14]   J. Peng, F. Zhang, G. Yan, Z. Qiu, and X. Dai, "Experimental study on rock-like materials
        fragmentation by electric explosion method under high stress condition", *Powder Technol.,* vol. 356,
        pp. 750-758, 2019.
        [http://dx.doi.org/10.1016/j.powtec.2019.09.001]

[15]   A.T. Odularu, "Metal Nanoparticles: Thermal Decomposition, Biomedicinal Applications to Cancer
        Treatment, and Future Perspectives", *Bioinorg. Chem. Appl.,* vol. 2018, pp. 1-6, 2018.
        [http://dx.doi.org/10.1155/2018/9354708] [PMID: 29849542]

[16]   C.N. Cheaburu-Yilmaz, H.Y. Karasulu, and O. Yilmaz, "Chapter 1 - Nanoscaled dispersed systems
        used in drug-delivery applications", In: *Micro and Nano Technologies* Elsevier Inc., 2018, pp. 437-
        468.
        [http://dx.doi.org/10.1016/B978-0-12-813932-5.00013-3]

[17]   R. Jose Varghese, E.H.M. Sakho, S. Parani, S. Thomas, O.S. Oluwafemi, and J. Wu, "Chapter 1 -
        Introduction to nanomaterials: synthesis and applications", In: *Nanomaterials for Solar Cell
        Applications* Elsevier Inc., 2019, pp. 75-95.
        [http://dx.doi.org/10.1016/B978-0-12-813337-8.00003-5]

[18]   M.A. Virji, and A.B. Stefaniak, "Chapter 1: A Review of Engineered Nanomaterial Manufacturing
        Processes and Associated Exposure", In: *Comprehensive Materials Processing, Health Safety and
        Environmental Issues* vol. 8. Elsevier, 2014, pp. 1-3.
        [http://dx.doi.org/10.1016/B978-0-08-096532-1.00811-6]

[19]   D.M. Mattox, *Physical Sputtering and Sputter Deposition.* Sputtering, 2010.

[http://dx.doi.org/10.1016/B978-0-8155-2037-5.00007-1]

[20] A.H. Simon, "Chapter 1 - Sputter Processing", In: *Handbook of Thin Film Deposition (Fourth Edition)* Elsevier Inc., 2018, pp. 195-230.
[http://dx.doi.org/10.1016/B978-0-12-812311-9.00007-4]

[21] A. Bashir, T.I. Awan, A. Tehseen, M.B. Tahir, and M. Ijaz, "Interfaces and surfaces", In: *Chemistry of Nanomaterials.*, 2020, pp. 51-87.
[http://dx.doi.org/10.1016/B978-0-12-818908-5.00003-2]

[22] M. Wang, and Q. Zhao, "Electrospinning and Electrospray for Biomedical Applications", In: *Reference Module in Biomedical Sciences* vol. 1–3. Elsevier Inc., 2019.
[http://dx.doi.org/10.1016/B978-0-12-801238-3.11028-1]

[23] M.T. Reetz, M. Winter, R. Breinbauer, T. Thurn-Albrecht, and W. Vogel, "Size-selective electrochemical preparation of surfactant-stabilized Pd-, Ni- and Pt/Pd colloids", *Chemistry,* vol. 7, no. 5, pp. 1084-1094, 2001.
[http://dx.doi.org/10.1002/1521-3765(20010302)7:5<1084::AID-CHEM1084>3.0.CO;2-J]   [PMID: 11303867]

[24] F. Bensebaa, "Wet Production Methods", *Interface Science and Technology,* vol. 19, pp. 85-146, 2013.
[http://dx.doi.org/10.1016/B978-0-12-369550-5.00002-1]

[25] C.G. Granqvist, and R.A. Buhrman, "Ultrafine metal particles", *J. Appl. Phys.,* vol. 47, no. 5, pp. 2200-2219, 1976.
[http://dx.doi.org/10.1063/1.322870]

[26] M.J. Luton, C.S. Jayanth, M.M. Disko, S. Matras, and J. Vallone, "Multicomponent Ultrafine Microstructures", *Proc. MRS,* vol. 132, p. 79, 1989.
[http://dx.doi.org/10.1557/PROC-132-79]

[27] C. Suryanarayana, and B. Prabhu, *Synthesis of Nanostructured Materials by Inert-Gas Condensation Methods.* 2nd ed. Nanostructured Mater. Process. Prop. Appl, 2006, pp. 47-90.
[http://dx.doi.org/10.1016/B978-081551534-0.50004-X]

[28] S. Barua, X. Geng, and B. Chen, "Chapter 3 - Graphene-based nanomaterials for healthcare applications", In: *Photonanotechnology for Therapeutics and Imaging* Elsevier Inc., 2020, pp. 45-81.
[http://dx.doi.org/10.1016/B978-0-12-817840-9.00003-5]

[29] J.P. Raval, P. Joshi, and D.R. Chejara, "Carbon nanotube for targeted drug delivery", In: *Applications of Nanocomposite Materials in Drug Delivery* Woodhead Publishing , 2018, pp. 203-216.
[http://dx.doi.org/10.1016/B978-0-12-813741-3.00009-1]

[30] J. Singh, T. Dutta, K.H. Kim, M. Rawat, P. Samddar, and P. Kumar, "Green synthesis of metals and their oxide nanoparticles: applications for environmental remediation", *J. Nanobiotechnology,* vol. 16, no. 1, p. 84, 2018.
[http://dx.doi.org/10.1186/s12951-018-0408-4] [PMID: 30373622]

[31] X. Li, H. Xu, Z.S. Chen, and G. Chen, "Biosynthesis of nanoparticles by microorganisms and their applications", *J. Nanomater.,* vol. 2011, pp. 1-16, 2011.
[http://dx.doi.org/10.1155/2011/270974]

[32] D. Zhang, X. Ma, Y. Gu, H. Huang, and G. Zhang, "Green synthesis of metallic nanoparticles and their potential applications to treat cancer", *Front Chem.,* vol. 8, p. 799, 2020.
[http://dx.doi.org/10.3389/fchem.2020.00799] [PMID: 33195027]

[33] J.K. Patra, and K. Baek, "Green nanobiotechnology: factors affecting synthesis and characterization techniques", *Journal of Nanomaterials,* vol. 2014, 2014.
[http://dx.doi.org/10.1155/2014/417305]

[34] M. Shah, D. Fawcett, S. Sharma, S.K. Tripathy, and G.E.J. Poinern, "Green synthesis of metallic nanoparticles *via* biological entities", *Materials (Basel).,* vol. 8, no. 11, pp. 7278-7308, 2015.
[http://dx.doi.org/10.3390/ma8115377]

[35]   Y.N. Slavin, J. Asnis, U.O. Häfeli, and H. Bach, "Metal nanoparticles: understanding the mechanisms behind antibacterial activity", *J. Nanobiotechnology,* vol. 15, no. 1, p. 65, 2017.
[http://dx.doi.org/10.1186/s12951-017-0308-z] [PMID: 28974225]

[36]   O.T. Fanoro, and O.S. Oluwafemi, "Bactericidal antibacterial mechanism of plant synthesized silver, gold and bimetallic nanoparticles", *Pharmaceutics,* vol. 12, no. 11, p. 1044, 2020.
[http://dx.doi.org/10.3390/pharmaceutics12111044] [PMID: 33143388]

CHAPTER 2

Analytical Methods in the Characterization of Green Nanomaterials

Abstract: A new class of diagnostic and therapeutic tools for various diseases has been made possible by advancements in polymeric nanoparticles as innovative nanomedicines. Although there are many benchtop studies in the nanoworld, their application to already marketed goods is still in its infancy. Problems with nanomedicine characterization cause this lack of transference, among other things. Three nanoscale characterization approaches may be distinguished: physicochemical property characterization, biological interactions of nanomaterials, and analytical characterization and purification procedures. Physical qualities may be assessed using a variety of methods in many situations. Choosing the best appropriate method is made more difficult by many advantages and disadvantages of each methodology; frequently, a combinatorial characterization approach is required. Scientists from many domains must find answers to the difficulties in reliable characterization of the nanomaterials after their fabrication and various systematic stages.

Keywords: Analytical methods, Characterization techniques, Green nanomaterials.

INTRODUCTION

Since NPs have a vast surface ratio compared to their bulk counterparts, nanoscale materials have a fast-increasing molecular reactivity. These include mechanical characteristics that can vary widely amongst nanoparticles (NPs) and chemical, optical, and electrical properties. Such nanostructures can be created *via* various techniques, including mechanical, chemical, and other approaches. More nanomaterials are being created now than ten years ago, necessitating the creation of more accurate and reliable techniques for their characterization. Such characterization can, however, occasionally be lacking. This is because nanoscale materials are more difficult to analyze adequately than bulk materials. According to the complex basis of nanoscience and nanotechnology, not every research team can easily access distinct characterization techniques. Sometimes it is necessary to describe NPs in a broader sense, which necessitates a comprehensive approach that incorporates complementary approaches. The chapter provides several novel ways for characterizing NPs about the qualities being studied (Tables **1** and **2**).

Seyed Morteza Naghib and Hamid Reza Garshasbi

Table 1. Parameters and corresponding characterization techniques.

Entity Characterized	Characterization Techniques Suitable
Size	TEM, XRD, DLS, NTA, SAXS, SEM, EXAFS, ICP-MS, UV-Vis, MALDI, NMR
Shape	TEM
Elemental-chemical composition	XRD, XPS, ICP-MS, SEM-EDX, NMR
Crystal structure	XRD, EXAFS, STEM
Size distribution	DLS, DTA, ICP-MS, NTA, SAXS, SEM
Chemical state–oxidation state	XAS, EELS, XPS
Growth kinetics	SAXS, NMR, TEM
Surface area, specific surface area	BET
Surface charge	EPMA, Zeta potential
Concentration	ICP-MS, UV-Vis
Agglomeration state	Zeta potential, DLS, UV-Vis, SEM, TEM
3D visualization	SEM
Detection of NPs	TEM, SEM, STEM

Table 2. Summary of the experimental techniques.

Technique	Main Information Derived
XRD	Crystal structure, composition, crystalline grain size
XAS (EXAFS, XANES)	X-ray absorption coefficient – chemical state of species, interatomic distances, Debye-Waller factors
SAXS	Particle size, size distribution, growth kinetics
XPS	Electronic structure, elemental composition, oxidation states, ligand binding
FTIR	Surface composition, ligand binding
NMR	Ligand density and arrangement, electronic core structure, atomic composition, NP size
BET	Surface area
TGA	Mass and composition of stabilizers
UV-Vis	Optical properties, size, concentration, agglomeration state
PL spectroscopy	Optical properties, size, composition
DLS	Hydrodynamic size, detection of agglomerates
NTA	NP size and size distribution
ICP-MS	Elemental composition, size, size distribution, NP concentration
TEM	NP size, size mono dispersity, shape, aggregation state, study growth kinetics

(Table 2) cont.....

Technique	Main Information Derived
STEM	Combined with HAADF, EDX for morphology study, crystal structure, and Elemental composition.
EELS(EELS-STEM)	Type and quantity of atoms present, chemical state of atoms, collective interactions of atoms with neighbors, bulk plasmon resonance
SEM-HRSEM, T-SEM-EDX	Morphology, dispersion of NPs in cells and other matrices/supports, precision in lateral dimensions of NPs, quick examination–elemental composition

ANALYTICAL METHODS

Size and shape are two of the most important characteristics to consider while determining the NP's identity. Size distributions, aggregation levels, surface charges, and surface areas can all be assessed as the chemistry on the surface. Its additional properties and applications may be influenced by the NPs' size, distribution, and organic ligands on their surfaces. Currently, there are no clear guidelines in place for this. Measurements based on NP can considerably impact whether these materials are accepted for commercial use while ensuring compliance with regulatory requirements.

Notwithstanding these issues, nanomaterials analysis remains problematic because of their multidisciplinary nature, lack of appropriate reference materials, challenges with specimen processing for analysis, and difficulty in data interpretation. The characterization of NPs has numerous obstacles, including the inability to measure in-situ and online concentrations of NPs, particularly in large-scale production, or to analyze NPs embedded in complex matrices. Large-scale production waste and effluent need to be closely monitored as well. As the production of NPs increases in volume, more accurate measurement methods will be required. As a result, characterizing NPs made in diverse methods is essential. We study the surface ligands and the nanoparticle core to understand how they affect the particle's physical properties. In addition, we do not just focus on the most prevalent methods. The kinetics of nanoparticle creation may be monitored using modern in situ operando techniques. Recent advances in controlled defects considerably impact nanoparticle properties [1].

OPTICAL CHARACTERIZATION TECHNIQUES

Confocal Laser-Scanning Microscopy

Confocal laser scanning microscopy (CLSM) is a potent method for producing clear pictures of a sample that might otherwise seem blurry when seen under a normal microscope. By capturing several photographs at various depths within a thick object, it is feasible to reconstruct 3D structures from the images produced using this approach. Confocal microscopy creates images by scanning a material with one or more focused beams of light. After focusing the laser beam using an objective lens on the specimen, the item is scanned with computer-controlled scanning equipment. A photomultiplier tube (PMT) detects the successions of light points on the specimen, but the PMT yield is combined into an image and shown by the system [2].

Scanning Near-Field Optical Microscopy

Scanning near-field optical microscopy (SNOM), which concurrently assesses topography and optical properties, establishes a direct relationship between surface nanofeatures and optical/electronic attributes (fluorescence). The fundamental idea of SNOM is to scan a sample surface with an arbitrarily small aperture, lit from behind at a close but constant distance, and record optical data pixel by pixel, adding transmitted, reflected, or fluorescent light to form a picture. In SNOM, an aperture with a diameter smaller than the excitation wavelength is used to concentrate the excitation laser light, resulting in an evanescent field in the contrary direction of the aperture. The sole element limiting the optical resolution of light traveling through or reflecting from the sample when scanned closely below the aperture is the aperture's diameter [3]. With a resolution less than the diffraction limit, this technique is excellent for quickly and conveniently imaging a material's optical characteristics.

It can be applied to the optical detection of a small surface in various fields, including materials research, nanotechnology research, nanophotonic and nano-optics research, and medical sciences. This approach makes it simple to identify single molecules. Dynamic characteristics can also be investigated at the subwavelength level.

Two-Photon Fluorescence Microscopy

Two-photon laser scanning microscopy, which allows both *in vivo* and in situ imaging, is well-established for researching biological systems. Denk, Webb, and collaborators devised this approach in 1990 [4]. In the nonlinear approach of two-photon fluorescence excitation of molecules, two photons must be absorbed whose total energy is more than the energy difference between the excited and ground states of the molecule and sufficient to affect a chemical shift to an excited electronic state. Because this method is reliant on it, the likelihood of a fluorescent molecule absorbing two infrared photons at the same time is a quadratic function of excitation radiance. In physiology, neurology, embryology, and tissue engineering, two-photon microscopy is used to observe highly scattering tissue. Two-photon microscopy might be used in a noninvasive optical biopsy that necessitates fast imaging.

Dynamic Light Scattering

Dynamic light scattering (DLS) is a noninvasive technique for assessing the size of particles and molecules in suspension [5]. Temperature, sample viscosity, and particle size influence the pace at which particles move in a Brownian motion. The particles were dispersed throughout the medium as a consequence of this motion. When one monochromatic light source, like a laser, interacts with a solution containing Brownian motion particles, the wavelength of the incoming light changes (Doppler shift). This shift in wavelength is correlated with particle size.

Unlike static light scattering, which monitors scattered intensity as a function of angle, DLS records the variation in scattered intensity over time at a fixed scattering angle (usually 90 degrees). Some incident light is dispersed in addition to the light scattered by a molecule or particle. If the molecule remained constant, the light's dispersion would not change. Moreover, as a consequence of interference, all molecules in the mixture disperse in a Brownian motion around the detector, altering the intensity of the light. Measurement of the period of light intensity variations by DLS can provide data on the normal size, size distribution, and polydispersity of molecules and particles in solution [6].

Brewster Angle Microscopy

A Brewster angle microscope (BAM) is used to view thin layers on liquid surfaces. This 1991 microscopy approach allows direct viewing of extremely thin organic coatings on transparent dielectric surfaces. Applying p-polarized light under Brewster's angle incidence results in no reflectivity from air-water interaction. Under a constant angle of incidence, a monolayer on the water's surface alters the Brewster angle condition, causing light to reflect [7]. This is the main method of providing contrast for a surface covered with a nanofilm in BAM. When certain light beams cross the boundary between two media with different refractive indices, they frequently reflect back. Light with a given polarization state cannot be reflected at a specified angle of incidence known as the Brewster angle (qB). This polarization is characterized as P-polarized light. BAM could scan dispersed or adsorbed monolayers.

BAM is used to examine phase transitions, specify domain microstructure, and phase-split mixed monolayers, track changes brought on by complicated evolution, and analyze phase transitions [8].

ELECTRON PROBE CHARACTERIZATION TECHNIQUES

Scanning Electron Microscopes (SEM)

This imaging technique is called scanning electron microscopy, or SEM for short (SEM). An electron beam is moved across a sample's surface in a specific pattern known as a raster scan. It has been shown that the released electrons might either be secondary or backscattered. Field emitter SEM (FE-SEM) uses a cathode to emit electrons when subjected to a strong electric field. The electron beam may destroy biological material, notwithstanding the ease of sample preparation.

Consequently, it is only appropriate to utilize it in small samples. In an environmental/wet SEM technique, wet SEM does not dry the sample, preserving its topography. SEM is an effective tool for assessing the NMs purity and degree of aggregation [9].

Whereas SEM offers NP Morphology information, EDS estimates sample composition. One may manually gauge the size of certain particles. An electron beam with a relatively low energy level (1-30 keV) scans the material's surface in the SEM to provide a picture with a resolution of a few nanometers. Due to the need to dehydrate and coat samples with gold before microscopy, conventional electron microscopes are also high-vacuum equipment that might change samples. In various ways, environmental SEM equipment contributes to this effort by enabling the investigation of wet materials. However, they sacrifice clarity in order to do so. Segmenting solid specimens for 3D imaging using focused ion-beam SEM and cryo-SEM techniques is possible. While real water samples with turbidity and colloidal characteristics can benefit from liquid-SEM- and environmental-SEM specialized techniques (at the cost of some resolution loss) [10]. Increasing the resolution of current electron microscopy, as well as acquiring more information about nanomaterials, has been promoted by nanotechnology.

It is a technique that may be used in various contexts, is non-destructive and gives substantial information on the structure, arrangement, and composition of nanomaterials. High-quality pictures and information regarding nanomaterial composition may be obtained using FESEM, which is necessary for many demanding nanotechnology applications. Nanomaterials and NPs surface morphology have been investigated using SEM. The size of nanomaterials, which may vary from hundreds of nanometers down to the atomic scale, is a major determinant of their properties (about 0.2 nm). For example, nanomaterials can have different chemical and physical properties than their bulk counterparts. Thus, nanomaterials have an improved ability to react with chemicals because of their higher specific surface area. Photocatalysis and electrocatalysis are two types of catalysis that profit from the high reactivity of nanomaterials.

Quantum events may dramatically alter a material's optical, magnetic, and electrical characteristics. Changing the color of gold NPs is possible based on their size. Creating nanomaterials with consistent size distribution and uniform growth is a major challenge in nanoscience. Specific applications necessitate the creation of NPs of the same size. As a result, one of the most important issues is to create NPs of uniform size. It is also difficult to manage and develop uniformly shaped nanomaterials. Shape management of nanomaterials has garnered interest because it allows for more exact tweaking of characteristics than ordinary NPs. Metal NPs with diameters less than 10 nm may include cuboctahedrons and octahedrons. This is because the crystallographic surfaces, which differ in the surface atom concentration and the surface energy, are responsible for forming these structures. This is because these surfaces have different crystallographic orientations.

Twinned metal particles are also discovered. Twinning occurs when two sub-grains share the same crystallographic plane. Multiple twinning occurs on alternating coplanar planes (decahedron and icosahedron) to create cyclic polyhedrons.

Fivefold axes organize the twinned tetrahedral subunits. Future electronics and nanoscience are interested in nanomaterial superstructures (ordered assembly or superlattices). For two- and three-dimensional superlattice structures, metal NPs (gold or silver) are suitable "building blocks". Passivation of the surface of the building blocks is a common step in preparing ordered assemblies to protect their properties from the environment and prevent sintering.

NPs can self-assemble into superlattices because their organic surfaces are protected from the environment. It is possible to alter the properties of the nanoparticle superlattices by altering the chemical functionality of a natural monolayer that surrounds them. The FESEM offers high-resolution surface imaging in the field of nanoscience because it analyzes surface morphology, shape, features, size, composition, and crystalline structure using a stream of very powerful electrons. This is possible because of the FESEM. Discovering the surface structure of various materials may be accomplished using a FESEM. A High-resolution FESEM examination of the co-doped CuO/ZnO nanomaterials revealed a chemical sensor nanostructure component with nanorods form. Zinc oxide was responsible for rod morphology, while copper oxide was responsible for the aggregated rod morphologies. ZnO rods had a morphology of an average of 530 nm in diameter.

CuO produced the aggregated rod shape in eight m. These aggregated rods, however, also include 183 nm sub-nanorods. Dense ZnO NPs with flat surfaces and a methanol sensor were produced for environmental remediation. We recently synthesized coumarin using ZnO NPs based on cellulose acetate polymer, and FESEM imaging revealed the floral shape morphology. These nanomaterials feature flower-shaped morphologies but appear sheet-like when aggregation occurs.

Nanomaterials based on Co_3O_4 were produced by increasing particle morphologies of CeO_2/Co_3O_4, calcined CeO_2/Co_3O_4, TiO_2/Co_3O_4, and Fe_2O_3/Co_3O_4. These nanomaterials are used for electrochemical water splitting. After being imaged with FESEM, Co-NiAl takes the form of a sheet.

On the other hand, Co-Al and Co-Ni were created as spherical NPs. Fe_2O_3/Co_3O_4 was synthesized for electrochemical water splitting using particle and fiber morphologies. At the same time, the other Co-based materials were assessed as spherical particles. The Co_3O_4/SiO_2 showed mesoporous morphologies because of its large surface area. Many different dyes can be reduced by photocatalytic reduction of Zn-Al/C and Zn-Cr/C nano-catalyzed in double hydroxide structures. A sheet-like shape and structure were seen in the FESEM images of both catalysts. These sheets, on the other hand, were composed of minute particles. So, it may be deduced that the sheet-like form of these microscopic particles can be attributed to their aggregation.

Cd-based materials had their sheet morphology examined using FESEM in yet another way. Nanomaterials were used for Cd-Al/C and CdSn/C. However, both had sheet shapes and were produced in hydroxides. Because of its double hydroxide morphology, the Cd-AL/C succeeded more than the Cd-Sb/C in decolorizing organic dyes. $CoSb_2O_6$ was discovered under FESEM to have a homogenous form for chemical sensors and environmental remediation. It was also discovered that the nanosheets of $CuO-TiO_2$ were aggregated. 4-nitrophenol sensing was accomplished by creating an Ag_2O particle with a consistent diameter and shape. Pt NPs are used in oxidation of toluene, which is done using H-type ZSM-5-produced mesoporous core-shell composites. FESEM is necessary to identify and evaluate inorganic-organic interaction or dispersion. Polymer hosts regulate several properties of the synthesized materials by allowing inorganic components to be dispersed throughout the substance. FESEM is commonly used for characterization of polymers and composite materials .

FESEM images of polymer host materials revealed ZnO spots in poly (propylene carbonate)/ZnO composite composites. Over 5% ZnO content in the polymer host materials caused polymer host aggregation. Composite materials are often used in packaging. Symmetrically shaped materials made of polymers such as polyether sulfones and cellulose acetates were created. We produced chitosan polymer host materials for organic dye adsorption using Co_3O_4/SiO_2 adsorbent materials. Co_3O_4/SiO_2 materials may be seen as a white region in the FESEM pictures. PUA and TZnO-W composite films, on the other hand, exhibited consistent TZnO dispersion. Antibacterial activity was shown to be very high in these films for the adsorption of organic dyes. Also, the chitosan material was coated with Ni/Al nanostructured materials in another embodiment, which showed excellent antibacterial properties and high absorption capacity.

Spherical materials with antibacterial and nitrophenol adsorption properties were found in all of these samples, regardless of their shape. FESEM allowed researchers to observe the PES polymer's roughness, SiO_2, cellulose acetate, and carbon black in the polymer. PES-cellulose acetate carbon black composites were coated with Cu0 NPs in the last stage of the experiment. NPs of CuO are excellent in reducing 4-nitrophenol. The PESCA-Ag_2O membrane's contact angle, mechanical strength, and water permeability were examined. These membranes were coated with CuO NPs and then used to reduce 4-nitrophenols. Good antibacterial activity was found in all membranes and CuO nanoparticle-supported membranes.

FESEM pictures showed the agglomeration morphology of Ag_2O particles. After introducing Ag_2O NPs into the original materials, a uniform membrane was seen, in stark contrast to the thick and porous character of the originals. The uniformity of the Cu NPs in the presence of PES-CA and PES-CA-Ag_2O is apparent [11].

Scanning Probe Electron Microscopy

Scanning probe microscopes (SPM) are imaging devices that examine atoms, nanoscale surfaces, and structures. In SPM, which uses light waves for imaging, a very tiny probe, or "tip", probes the material's surface. Then, it is observed how strong the interactions are among the surface and the tip. A few angstroms or nanometers distant, an atomically sharp probe detects the surface to provide a three-dimensional topographic image of the area. A powerful kind of microscope with a resolution of under 1 nanometer is a SPM. Scientists frequently use one of the two scanning modes: contact mode, physical touch, or noncontact mode, depending on the kind of information they hope to learn from the study. The scientist may quickly take a surface shot since the energy on the point, and the region is constant in contact mode. When in tapping mode, the cantilever oscillates and momentarily touches the ground. Imaging a soft surface is better when done in the tapping mode. Scanning probe microscopes come in several different varieties. Electrostatic forces between the cantilever tip and the sample are tracked by atomic force microscopes. Although scanning tunneling microscopes keep an eye on the electrical current flowing among the cantilever tip and the sample, magnetic force microscopes assess magnetic forces. The topographical, electrical, and magnetic characteristics of the samples are all known. SPM may also gather information and record it on a material [12].

Electron Probe Microanalysis

X-ray spectroscopy is used in electron probe microanalysis (EPMA) to determine the chemical composition of a given area in multiphase materials. It is a non-destructive, in-situ chemical examination that delivers a comprehensive image of the material. Through this approach, the intensity of a substance's distinctive X-rays is investigated. Enough energy is released when an accelerated and focused electron beam strikes a solid object to liberate matter and energy from the sample.

An electron microprobe operates in this manner. These reactions between the electron specimens result in derivative electrons, X-rays, and heat emission. Secondary and backscattered electrons can be used to observe a region or determine the typical material composition while examining geological materials. After an inner-shell electron is kicked off of its orbit, a higher-shell electron must produce an X-ray to replace the empty space. X-rays are created by the interaction of the incoming electrons with the atoms' inner-shell electrons in the sample. These quantized X-rays serve to identify the element. As the sample volume is not lost due to the X-rays generated by the electron interactions, EPMA is categorized as "non-destructive," enabling several investigations of the same substance. The most used technique for chemically analyzing small-scale geological materials is quantitative EPMA. Most often, EPMA is used when it is necessary to investigate each phase separately; the material is tiny or important for other reasons. Sometimes it is possible to determine a mineral's age without doing isotopic ratio tests, such as monazite [12].

Transmission Electron Microscopy

A transmission electron microscope (TEM) uses strong electrons to gather information from materials, including morphological, chemical, and crystallographic data. TEM, which uses electrons rather than light, operates on the same fundamental principles as light. Since electrons have a smaller wavelength than light, TEM images have far greater resolution than light microscope images. As a result, transmission electron microscopes may reveal the minute details of internal structures, frequently down to the level of one atom. The employment of electromagnetic lenses rather than glass lenses and the viewing of pictures on a screen rather than through an eyepiece are two other differences between TEM and light microscopy. Direct observation of a material's atomic structure is possible using high-resolution [13].

Scanning Transmission Electron Microscopy

The scanning transmission electron microscope is a useful instrument for analyzing nanostructures because it provides a broad variety of imaging modes and has the capacity to provide information on the electrical structure and elemental composition at the high sensitivity of an atom. Scanning TEM (STEM) uses any piece of equipment to combine the ideas of TEM with scanning electron microscopy. STEM employs very thin samples and, like TEM, concentrates on the electron beam delivered by the specimen. Once an electron nanoprobe contacts a specimen within a STEM device, many electrons, electromagnetic signals, and other signals may be produced.

Pictures, diffraction patterns, or spectroscopic information about an object can be created using any of these signals. For instance, it is possible to use an annular detector to collect high-angle scattered electrons and produce images showing atomic-level structural reforms across the sample. On an atomic or sub-nanometer scale, electron energy-loss spectroscopy (EELS), which is based on the energy analysis of inelastically scattered electrons, can provide details on the electrical structure, oxidation states, and chemical composition. X-ray energy-dispersive spectroscopy (XEDS) can quantify differences in elemental composition brought on by atypical sample structures. It is possible to perform atomically precise and sensitive analyses of the composition, chemistry, electrical structure, and crystal structure of nanoscale systems by fusing XEDS and EELS with HAADF imaging. By gathering or examining secondary electron and Auger electron signals released from a specimen surface, we can learn about the surface topography or composition of the sample.

Using an electron nanoprobe in the target area, it is feasible to get coherence electron nano-diffraction patterns of particular nanocomponents. These patterns can provide a plethora of information about an object's nanostructure [14].

PHOTON PROBE CHARACTERIZATION TECHNIQUES

Photoelectron Spectroscopy

Photoelectron spectroscopy (PES) examines the energy of electrons released by the photoelectric effect from solids, gases, or liquids to estimate electron binding energies in a material [15]. The PES method uses the photoelectric effect as its underlying physics. This method has been split into two types based on the source of the exciting radiation.

Ultraviolet PES: Between 10 and 50 eV photon energy, greater than typical work function values (2-5 eV). As a result, the photoelectric effect allows electrons to be expelled from surfaces. The exploration of chemical bonding and valence energy levels, notably the bonding properties of molecular orbitals, uses UPS.

X-ray PES (XPS): An element's empirical formula, chemical state, electronic state, and part per thousand elemental compositions may all be determined using the surface-sensitive quantitative spectroscopic technique known as XPS. X-ray photoelectron spectra are created by exposing a substance to an X-ray beam and counting the quantity and kinetic energy of electrons, leaving the top 0 to 10 nm of the material under investigation. XPS, also known as electron spectroscopy for chemical analysis, evolved into a method for examining the surface chemistry of substances in both their untreated and treated states.

UV-Visible Spectroscopy

UV-Visible spectroscopy, which produces light with a wavelength ranging from 190 to 800 nm, may be used to determine the concentration of a compound and its size and form in certain situations. You can use it to analyze a wide range of nanomaterials.

It has been used, for example, to calculate the biomolecule-nanomedicine ratio. Because most materials only absorb a certain wavelength of light, interference between the absorption of distinct substances is considered.

Plasmonic NPs absorb radiations in the near-infrared range (NIR) based on size and structure. Known as SPR, this feature is connected to the collective oscillation of nanoparticle surface electrons. Using one or more peaks in the dispersion of NPs, it is possible to get information on their size, shape, and distribution. Microgels containing plasmonic NPs are often studied using UV/vis spectroscopy.

According to Farooqi *et al.*, UV/Vis spectroscopy was utilized to examine P(NIPAM-AA) microgels and hybrids of Ag and P (NIPAM-AA). Microgel dispersion samples were scanned in the 200–800 nm UV/vis range for this investigation. This band was found to be absent from all but the purest microgels. This study shows that hybrid microgels' dispersion is still restricted at 420 nm. Only after a single peak at the SPR wavelength (420 nm) could it be noticed was it determined that NPs had been effectively loaded into a polymer network. NPs width at half maximum has a fairly narrow fair value. Zhang and his colleagues employed UV/Vis spectroscopy to understand better the optical characteristics of CdS and poly (N-isopropyl acrylamide-acrylic acid-2-hydroxyethyl acrylate) hybrid microgels. Functional groups were produced by copolymerizing acrylic acid (AA) to couple metal ions. In order to concurrently increase the filter size, heac, a 2-hydroxy ethyl acrylate, was utilized. Nanoparticle size and polydispersity were assessed using UV/Vis spectroscopy.

For the hybrid microgels, in order to investigate the effects of different CdS and Ag NP concentrations, UV/vis spectroscopy was used. There were significant increases in NP. Absorption when CdS's content was increased from 0.027 to 0.08 grams. The hybrid microgels utilized in this work varied in size from 3.0 to 5.9 nanometers, depending on their composition. After 12 hours of reflux, samples of CdS-P (NIPAM-AA-HEAc) hybrid microgel was analyzed using UV/Vis spectroscopy.

The absorption peaks in the UV/Vis region are improved by heat treatment to control CdS NP polydispersity in microgels. This is the result of the Ostwald ripening of tiny NPs.

In addition, UV/Vis spectroscopy was used to investigate P(NIPAM-AA-HEAc) microgels; similarly, conclusive results were obtained. In hybrid microgels created with 0.23 g of silver NPs per gram of polymer, a significant absorption peak was seen at 411 nm. Silver NPs generated with 0.39 grams of silver NPs per kilogram of polymer, on the other hand, showed a prominent peak, increasing particle size due to the addition of additional silver NPs.

Au NRs have been incorporated into P(NIPAM-AA) microgels by Gorelikov *et al.* The hybrid system was studied using various UV/Vis spectroscopic methods. Two peaks in the dispersion of the hybrid microgels can be seen at 400 and 810 nm. These peaks are responsible for the transverse, and longitudinal SPR characteristics of the Au NRs inserted into the hybrid microgels. The appearance of two peaks in the dispersion spectrum of the hybrid system is evidence that nonspherical or rod-like NPs have been introduced into the microgels.

Furthermore, Lu and co-workers used UV/Vis spectroscopy to confirm that Ag NPs had been created in the shell of polystyrene-poly(N-isopropyl acrylamide) [PSTp(NIPAM) core-shell microgels. A single, narrow UV/Vis spectrum peak was used to load Ag NPs into the polymer network effectively. The size distribution of the Ag NPs was carefully regulated.

Microgels created by Suzuki *et al.* had a P(NIPAM) core encased in a P(NIPAM-APMa) shell that included functional groups that were positively charged. P was used in the construction of the second shell (NIPAM).

The chemical reduction process was applied to prepare Au seeds as part of the core-shell–shell microgel particle structure. After that, electroless plating from Au seeds was used to grow Au NPs.

Microgels containing Au seeds and NPs were examined in the UV/Vis range to determine their optical characteristics. Even though the dispersion of hybrid microgels became pink, the core-shell-shell microgel particles impregnated with au seed did not display a peak. Since Au seeds are so small (less than 4 nm), they failed to exhibit the SPR phenomenon.

Instead of creating new Au NPs, electroless plating was used to expand the desired size of Au NPs. There was an intrinsic peak at 520 nm for Au NPs. Gold NPs with a restricted distribution can be seen in the UV/Vis spectrum at 520 nm. At temperatures of 22 degrees Celsius, 38 degrees Celsius, and 43 degrees Celsius, Wu *et al.* generated silver-poly (NIPAMAA-AAm) hybrid microgels in the laboratory. They used an in-situ reduction procedure that was based on gold salt to achieve their goal of thoroughly saturating the materials with Au NPs. A room temperature examination was conducted at this stage. UV/Vis spectroscopy was used to study the microgels' UV/Vis spectroscopic properties for both types of microgels. It has been revealed that the 400 nm SPR band, which is critical to Ag NPs, has been erased. The Au NPs were 500 nm in diameter in each sample. The production of metal NPs in microgel particles has been studied by other researchers using UV/Vis spectroscopy.

Growth of NPs in Microgels

If the microgels are transparent, UV/vis spectroscopy may be used to monitor the formation of plasmonic NPs inside them. The surface band of the plasmonic NPs undergoes a redshift when they develop in microgels. P(NIPAM-AA) microgels were used by Kim *et al.* to generate gold NPs in situ. Another method for examining the growth of metal NPs in polymer shells is UV/Vis spectroscopy.

Contreras-Caceres *et al.* used cetyl trimethyl ammonium bromide to make core-shell microgels containing Au NPs (CTAB). Precipitation polymerization was used to build a P(NIPAM) shell around the center of the Au NPs.

At doses of 0.015 M to 0.05 M, CTAB was employed to study the influence on Au NPs' core shape. The UV spectra of two separate hybrid microgels showed the varied morphologies of the central Au NPs core. 0.05 M CTAB was used to generate a spherical-shaped Au NP core.

As the CTAB solution concentration was increased, a flower-like structure emerged, with random branches sprouting from the center in all directions. The SPR band was redshifted from 15 to 50 degrees Celsius when P(NIPAM) core-shell hybrid microgels were encased in an Au-NPs core-shell microgel. Additionally, the SPR peaks were more diffuse due to this process. The redshift was more pronounced in flower-like Au NPs core-enclosed P(NIPAM) core-shell hybrid microgels, and an extra band at 530 nm was seen. Au nano stars with a core-encased core-shell microgel core have been discovered to have two Plasmon modes, which resulted in two bands visible in the microgels. P(NIPAM-A--HEAc) microgels were made by Zhang *et al. via* precipitation polymerization. They made them utilizing Ag NPs and UV irradiation at 365 nm as a precursor to AgC ions from $AgNO_3$ salt. UV/Vis spectroscopy determined how long Ag NPs can develop within the polymer network.

The dispersion of microgels altered from light pink to purple to dark red as the irradiation period increased, showing the creation and development of Ag NPs in the system. Even before exposure to light, there was no peak in the UV/Vis spectrum at 300–700 nm when AgC-containing microgel dispersion was added. A 520 nm shoulder peak was noticed after three minutes of irradiation. The Plasmonic resonance phenomena of Ag NPs caused a peak at 440 nm after 6 minutes of irradiation. It was discovered that manufactured silver NPs had a higher absorbance intensity. "Poly(2-dimethylamino methacrylate-co-(trimethoxysilyl)-3-propyl, methacrylate) microgels" [P (MAEm-tMSPm)] Au NPs were studied using UV/Vis spectroscopy to see how they formed. 16 mL of 100 mgmL-1 $HAuCl_4$ solutions were added to a suspension of microgels and heated to 70°C. Spectra were scanned at 0, 5, 10, 15, 20, and 30 minutes. After 5 minutes of reaction progress, the SPR band of Au NPs developed in the 500–550 nm range and got more blue as reaction time advanced. Au NPs were formed after 30 minutes of ion reduction of gold. The optical characteristics of P(NIPAM-AAAAm) hybrid microgels made by Wu *et al.* at pH 8.48 3.33 were examined. Studies have been conducted at pH levels above and below pka. The pH and feed content of P(NIPAM-AA-AAm) microgels dictate that the volume phase transition temperature is 32–45°C. As a result, Ag NPs in microgels were inflated, partly shrunken, and completely collapsed by the in-situ reduction technique. The existence of Ag clusters was confirmed by the UV/vis spectra of Ag NPs, and NPs synthesized at all temperatures. Distinct Plasmonic band locations and shapes might be achieved at various medium temperatures. Because of the inflated microgel particles, large Ag NPs of various sizes were produced when the temperature was 22 degrees Celsius. At 38 degrees Celsius, partly inflated microgel particles produced Ag NPs with restricted dispersion. At 43 degrees Celsius, the microgel particles collapsed into their final shape, revealing a faint SPR band indicating silver NPs or silver clusters.

Growth of Polymer Network

UV/Vis spectroscopy can track the development of the polymeric network surrounding Plasmonic NPs. An increase in the medium refractive index around NPs results in a shift in the Plasmonic band to a higher wavelength as the thickness of the polymeric shell increases. To make Au spheres, Au decahedrons, and Au nanostars, Lopez and co-workers used layer-by-layer assembly with

CTAB and poly(vinylpyrrolidone) as the supporting materials, respectively. With Au spheres, Au stars, and decagons, they created shells of P(NIPAM). In order to investigate how temperature changes affect the optical characteristics of the sample, the researchers played around with the refractive index around the central Au cores. These gold spheres, decahedrons, and nanostars have 58, 96, and 115 nanometers, respectively.

P (NIPAM) shell-enclosed gold sphere, decahedron, and nanostar UV/Vis spectra showed SPR bands at 540-620-850 nanometers. The location of the SPR band is determined by the size of the central gold core, according to the research results.

Region Pastoriza-Santos *et al.* used a layer of Polystyrene Sulfonate (PSTs) and a silica shell to study how the shell's growth affects the longitudinal SPR of gold NPs in a shell. The optical properties of a hybrid system depend on how thick the polymer shell is. Because of the bigger Au core, the SPR band in the UV/Vis region redshifts, resulting in greater intensity and bandwidth. Due to increased system shell thickness, the transverse Plasmon band remained unchanged; nevertheless, the longitudinal Plasmon band became redshifted. For Au NRs, an increase in shell thickness resulted in a local refractive index, which led the Plasmon band to redshift. Fig. (**1**) shows that the longitudinal SPR value of Au NRs was unaffected by a shell thickness of more than 33 nm, as predicted.

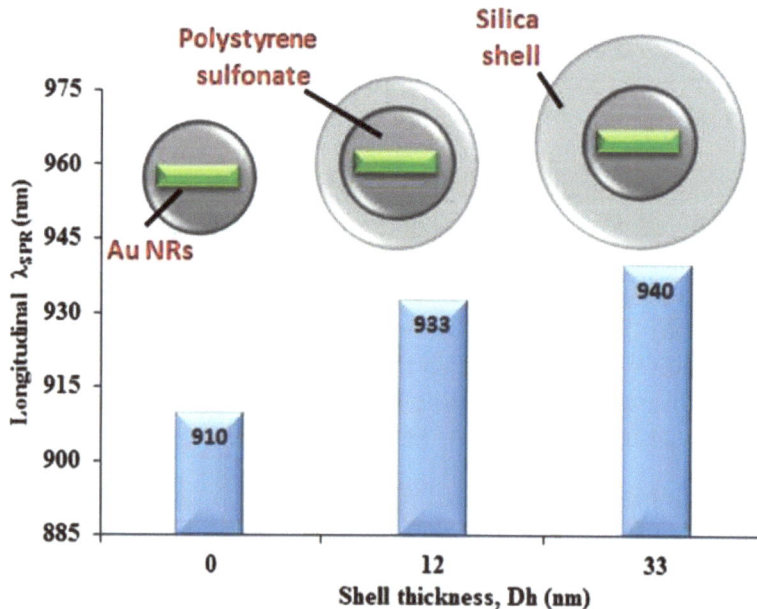

Fig. (1). SPR of Au NRs in an Au NRs-PSTs-silica composite as a function of increasing shell thickness [16].

Stability of Metal NPs in Microgels

According to published research, microgels have long-term stability owing to donor-acceptor interactions between polymeric network functional groups and metal NPs. UV/Vis spectroscopy may be used to verify the long-term stability of metal NPs in microgels. SPR wavelength measurements over time are used to conduct stability studies on NPs Hybrid microgel dispersions are kept at room temperature, and dark are monitored for changes in UV/Vis spectra.

Metal NPs in a polymeric network are stable if their λSPR value does not change. Changes have been made to the SPR value. Using UV/Vis spectroscopy, we have confirmed the stability of Ag NPs in P(NIPAM-AAm) microgels. Aluminum foil was used to seal a vial containing a diluted dispersion of p (NIPAM-AAm) microgels.

When P(NIPAM-AA) microgels were microwaved with $AgNO_3$ salt as a precursor to the synthesis of silver ions, the creation of Ag NPs was observed. Its UV/Vis spectra were studied up to six months after production. This indicates that Ag NPs may be utilized long in real applications since the λSPR value did not change. In the UV/Vis range, Ag NPs peaked at 429 nm due to their λSPR characteristics. Microgel particles containing Ag NPs were investigated for their stability. Following their preparation, Ag NPs were stable for eight months, according to the researchers. In order to investigate the stability of Ag NPs, UV/vis spectroscopy was used. The peak's absorbance intensity at 429 nm decreased marginally, but the SPR stayed the same. Ag-P(Nipamaa) hybrid microgels, on the other hand, retained their original hue. There seems to have been an extremely strong link between Ag NPs and microgel functional groups, which helped prolong their longevity.

Additionally, Wu and co-workers used UV/Vis spectrophotometry to scan the spectra of freshly generated samples and those in storage for one month to examine Ag-P(NIPAM-AA-AAm) hybrid microgels.

The UV/Vis spectra of Ag P(NIPAMMAEm) microgels [Ag P(NIPAMMAEm)] showed that silver nanoparticles (AgNPs) were very stable in microgel sieves even after eight months of manufacturing. It was discovered that the stability of silver NPs might be increased by including amide and carboxylic acid chelating groups into microgel templates. As discovered by Tang and colleagues, the reduction of methylene blue (MB) in these hybrid systems was not affected by 40 days of preparation.

Stability of Metal NPs Loaded Microgels at Different pH

In order to store metal nanoparticle-fabricated microgels for an extended time, the medium's pH must be controlled. When tested at various pH levels, the hybrid polymer microgels failed. The functionalities of the polymeric network dictate the pH range in which stable hybrid microgels may exist. In work by Farooqi *et al.*, poly (N-isopropyl acrylamide-methacrylic acid) microgel-fabricated Ag NPs were tested to see how well they held up in both acidic and basic environments. The microgels containing the Ag-P (NIPAM-Ma) hybrid had a pH of 9.9 (very basic), while the microgels serving as a control had a pH of 2.83. (Low, acidic). UV/Vis spectroscopy was utilized to examine the stability of microgels containing silver NPs in the previously mentioned medium. The UV spectra of the Ag-P(NIPAM-Ma) hybrid microgels were examined following the pH adjustment for 18 hours. The Ag NP SPR band at pH 9.9 did not shift or lose strength even after 18 hours. After 18 hours of incubation at pH 2.83 with Ag-P(NIPAM-Ma) hybrid microgels, the 400 nm SPR band disappeared. Microgels containing P(NIPAM-Ma) deprotonated Ma carboxylic acid groups at high pH of 9.9.

These negatively charged groups caused the polymer network to expand significantly, maintaining the stability of the Ag NPs. Increased polymer-polymer interactions resulted from the protonation of carboxylate groups at a low pH (2.83).

It was found that microgel particles in acidic pH values had shrunk, whereas those in neutral pH values had aggregated. These conditions led to the accumulation of AgNPs. As a result of the microgel particles' tendency to aggregate and shrink, we can no longer consider them to be in the nano range. This demonstrates that Ag-P(NIPAM-Ma) hybrid microgels do not exhibit a peak in the UV/VIS region when the pH of the solution is 2.38. P(NIPAM-Ma) microgels were able to preserve the stability of Ag NP at pH pka over a prolonged time. In an acidic environment, these microgels were exceedingly unstable.

Optical Properties of NPs Loaded in Microgels

Plasmonic NPs placed in responsive polymer microgels have a unique optical feature known as the SPR wavelength (λSPR). Its levels might fluctuate widely depending on the medium's temperature and pH. Changes in pH and temperature can affect the distance between NPs and the refractive index of the medium around the NPs, which can cause the microgel particles to either expand or contract. The SPR's value is impacted by it. Using UV/Vis spectroscopy, metal NPs integrated into microgels may be examined.

Effect of pH on Optical Properties of Metal NPs

NPs containing plasmonic metal NPs may have their optical characteristics, such as λSPR, modified by the pH of the media. Researchers from Farooqi and colleagues looked at how pH affected the surface plasmon resonance (SPR) of silver NPs embedded in P(NIPAMMa) microgels. They discovered that the medium's pH provided a significant effect. Table **3** displays the SPR values obtained for silver NPs generated in P(NIPAM-Ma) microgels at medium pH levels. The pH of the media rose from 3.27 to 9.90, increasing the spectral redshift of λSPR. When the pH of the medium rose, the SPR band's intensity decreased.

P (NIPAM-Ma) micelles deprotonated at high pH, shifting the SPR band's location by two when the pH was increased. The microgel network swelled as a result of the deprotonation of Ma groups. More water flowed into the polymer network as the microgel network swelled. A microgel network with Ag NPs was constructed, and the distance between the NPs increased.

Consequently, the electron density on the Ag NP surface was reduced. Electron oscillations were also reduced. The long-wavelength electromagnetic radiation in the UV/Vis range resonates with slow electron oscillation. As a result, Ag NPs' SPR band was visible at a larger wavelength.

Consequently, the 'SPR' of Ag NPs synthesized in P (NIPAM-Ma) microgels varied with the pH of the medium. The redshift in their λSPR value may not be solely due to increased inter-nanoparticle distance. Suppose the shift is due to a change in the spacing between the NPs or aggregation of some NPs due to microgel swelling. In that case, the investigations need to be done in reverse order. Variations in pH have the potential to have varied impacts on the refractive index of microgels that contain Ag NPs and are thus susceptible to these changes. Using UV/vis spectroscopy, several research teams have looked at the pH-dependent changes in hybrid microgels' optical properties [17].

Table 3. Effect of pH of the medium on λSPR of Ag NPs.

pH	λ_{SPR} (nm)
3.27	405
6.08	410
8.38	415
8.97	418
9.90	420

DRS Analysis

Known as DRS, differential reflectometry (DRS) is a surface analysis method. In order to gather information about the topic, it uses photons. The initial 10–20 nm of photon energy interacts with materials such as metals, alloys, and semiconductors. This technique allows DRS to reach 50–100 atomic layers deep into opaque materials. Surface techniques such as ESCA and DRS may penetrate up to 1, 5, or even 20 monolayers into the bulk material, whereas XRD can penetrate up to 50 micrometers. DRS bridges the gap between these two methods." The information gathered by DRS is distinct from that gleaned by the surface approaches outlined above.

Instead of scanning only one sample, DRS examines two with slightly different attributes (such as the composition of an alloy). A photomultiplier tube or other light-sensitive device receives the light reflected from this sample pair (PMT). After electronic processing, the resulting spectroscopic 'differential reflect gram' has all optical absorption spectrum characteristics. It is shown how electrons absorb photon energy as they transition from a lower, full state to a higher, empty state in the energy spectrum by using differential reflects grams to get the best results; it is best to use X-ray emission or Auger.

In the later spectra, electrons move between inner electron levels, resulting in new spectra. Optical spectra, on the other hand, are the consequence of electron transitions that occur between large outer electron bands. Optical spectra, or to put it another way, optical technologies such as DRS that exploit photon interactions with valence electrons rather than core electrons, offer information on the electrical structure at the Fermi surface. Another way to say this is that optical spectra provide this information. Electron interbond transitions are used to identify materials because each substance has its unique electron band structure.

To learn more about these energy levels, researchers can experiment with different solute components, transformations, defects in the lattice, ordering, and even ion-implantation surface states. DRS is not limited to materials with high absorption, such as metallic alloys and semiconductors. It has been shown that thin-film corrosion products on metal substrates can recognize and describe surface layers that are either transparent or semi-transparent. Due to its difference-forming nature, DRS has a significant benefit over conventional optical methods in that it may eliminate any unwanted effects on a differential reflect gram. As a result, the approach has an unusually high signal-to-noise ratio. Except for measurements in vacuum UV, no vacuum is required.

An automated scan from ultraviolet to visible to infrared, called a differential to reflect gram, takes around three and a half minutes to complete. CCD (charge-coupled device) cameras can take images in less than one second using a continuous ('white') light source and charge-coupled devices. It is, therefore, possible to see the formation of a surface layer owing to the interactions between organisms and the surrounding environment.

THE INSTRUMENT

Optics and Electronics

Two samples are placed side by side with almost little space between them. The differential reflection spectrometer measures the normalized difference in their reflectivity. Depending on the specifics of the case, the sample setup will vary. Two somewhat different specimens may be selected as an alternative.

With an oscillating mirror, unpolarized, monochromatic light (illuminated by, for example, a high-pressure xenon source) with a continuously variable wavelength is alternately deflected to one or both samples in a double monochromator (Fig. **2**). This mirror has a handy frequency of 50 or 60 Hz, making it easy to use. When the two samples' boundaries separate, the light beam's cross-section is rectangular and parallel. The beam passes across the specimens perpendicularly to the border. A light source with an incidence close to normal is used to illuminate the sample pair. These specimens are in sight of the monochromator's exit slit. The vibrating mirror that has already been placed on the ground plate formed of fused silica in front of the face of the PMT is amplified by this component thanks to its location in front of the PMT. Through a process known as plate diffusion, light falling on the PMT is softened and less likely to be affected by surface variations insensitivity, which can occur when a mirror is oscillating. An oscillating 60-Hz square wave modulates the PMT's DC signal.

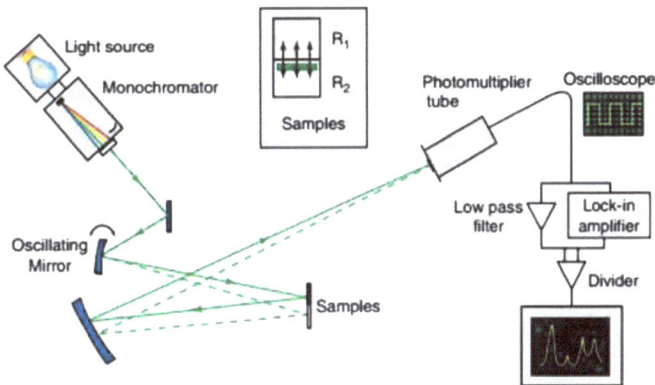

Fig. (2). Schematic diagram of the differential reflectometer [18].

A potentiometer coupled with a monochromator scanning gear supplies DC voltage proportionate to the light's wavelength. The recorder's X input is connected to this signal (or a computer). A differential reflects gram is generated automatically by plotting R/R vs. wavelength. Between 200 and 800 nm (1.6 to 6 eV), the scan takes around 1–3 minutes to complete. Better than 0.01% sensitivity to the normalized reflectivity difference. Line voltage variations are eliminated by measuring R1 and R2 simultaneously and calculating the R/R ratio. Spectral fluctuations in light output, detector sensitivity, and reflectivity of mirrors are also eliminated by this method. It analyzes the two samples using an oscilloscope linked to the PMT's output port. This technique may directly measure the normalized discrepancy in reflectivity between two samples. It tells you if the laser beam rests on each specimen simultaneously. By utilizing a difference-forming approach, any surface flaws that exist on either sample piece will be completely removed.

The sample can be moved up or down on the holder to fix errors. To sum up, DRS is fast, precise, and non-destructive. For example, it does not require polarized light or a vacuum to generate a continuous absorption-type spectrum. It removes surface disturbances that occur on both specimen parts. Under conditions of near-normal incidence, the data can be recorded. A personal computer has enhanced the analog instrument's digital components (PC).

Sample Holder

The solid samples are attached to a stage using double-sided tape or clamps so that they may be moved vertically and horizontally. An area of the sample's surface may be selected using this method. Additional advantages include the ability to orient both specimens on the scanning beam. Samples with greater reflectivity at one particular wavelength are put up to induce an R/R peak in an optical diffrometer's circuit. There is always a higher concentration of solute-containing alloy on top [18].

Materials' bandgaps are critical in concluding their photoactivity and conductivity. Bandgaps of nanomaterials can be determined using this method to their fullest extent. Metal-free water-splitting photocatalyst (C_3N_4) was discovered in carbon nanodots (C_3N_4). The band gap value was determined to be 2.74–2.77 eV using UV–Vis spectroscopy, which directly corresponds to the material's photo-ability. Doping or creating composites, as well as heterostructure NPs, may be studied using this method. MMT, $LaFeO_3$, and $LaFeO_3$/MMT nanocomposites were created by Peng *et al.* They used UV–vis DRS to study their electromagnetic radiation absorption fluctuation to understand their optical properties better. While pure MMT and $LaFeO_3$ NPs had smaller redshifts, the nanocomposite had a significantly larger one. Wide absorption bands from 400 to 620 nm reduced the $LaFeO_3$ and $LaFeO_3$/MMT bandgap.

The recombination and half-life of charge in the material conductance band affect all photo- and imaging-related applications. These photocatalysts are significant in solar light-driven chemistry because of this feature. Photoactive NPs and other nanomaterials can also be studied using photoluminescence (PL). Materials' absorption or emission capacity and their influence on photoexcitation time may be measured using this approach. Emission or absorbance measurements might be made depending on the purpose of the investigation. The decrease in charge recombination rate and the longer duration of photoexcited in the latter situation is to blame for the quenching of pure ZnO to CdS/Au/ZnO. NPs defects/oxygen vacancies can also be determined using this method, the thickness of a layer and the doping quantity of the material. Wan *et al.* used spectroscopic ellipsometry to determine the HGNPs' refractive index and extinction coefficient.

The optical constants of HG-NPs with varied morphologies and plasmonic characteristics were computed. Because of the materials' high sensitivity, which was revealed by measuring their ellipsometric radius, the results were compared to those of solid gold NPs with optical constants [19].

Inductively Coupled Plasma Spectroscopy

Using non-interfered, low-background isotopes, and inductively coupled plasma mass spectrometry (ICP-MS), metals and a number of nonmetals may be found at low concentrations such as 1 part in 1015 (parts per quadrillion). When the material has been ionized with ICP, the ions are extracted and quantified using a mass spectrometer. Atomic absorption spectroscopy is slower, less precise, and less sensitive than ICP-MS. In contrast to other MS techniques, such as thermal ionization MS and glow discharge MS, ICP-MS introduces a number of interfering species, such as argon from the plasma, airborne component gases that leak through the cone orifices, glassware contamination, and contamination from the cones themselves [20].

Fluorescence Spectroscopy

Electromagnetic spectroscopy, called fluorescence spectroscopy, examines fluorescence from a sample. Absorption spectroscopy is a supplementary method. Fluorescence and phosphorescence, two processes that result in photons from electrically excited states, occur during molecular relaxation. Transitions among the electronic and vibrational states of polyatomic fluorescent substances are a part of these photonic processes (fluorophores). The excited state structure and pertinent transitions may be seen in the Jablonski diagram, a useful illustration.

The transition from the ground state to the excited state occurs relatively quickly during the excitation process. Upon excitation, the molecule immediately relaxes to the excited electronic state's lowest vibrational level.

On time, range ranging from femtoseconds to picoseconds, this quick vibrational relaxation process takes place. Fluorescence emission happens when a fluorophore switches from a permitted vibrational level in the electronic ground state to a singlet electronic excited state. The fluorescence excitation and emission spectra, which represent the ground and excited electronic states, respectively, reflect the vibrational level structures [21]. It is an extensively applied and successful technique for various forensic, environmental, industrial, medical diagnostic, and biotechnology applications. It is a practical analytical technique that may be applied to quantitative and qualitative evaluations.

ION PARTICLE PROBE CHARACTERIZATION TECHNIQUES

Rutherford Backscattering

Rutherford backscattering spectrometry (RBS) is an ion scattering method used to analyze the composition of thin films. Being able to quantify without using reference standards makes RBS special.

The material is attacked with high-energy He^{2+} ions in an RBS experiment, and the transfer of energy and yield of the backscattered ions are then measured at a preset angle. Since the backscattering cross-section of each element is known, a quantitative compositional depth profile can be determined from RBS spectra acquired for films less than 1 mm thick. Examples of uses include identifying the composition and degree of contamination of surrounding films and silicide layers and also the thickness density of both [22].

Small-angle Scattering

The scattering method known as small-angle scattering (SAS) relies on the deviation of collimated radiation off its straight route when it interacts with objects considerably bigger than the radiation's wavelength. The term "small-angle" refers to the deflection, which is minimal (0.1–10 degrees). SAS techniques can determine a sample's structures' size, shape, and orientation. SAS may be used to efficiently study large-scale structures with various sizes between tens of angstroms to thousands or even tens of thousands. Its ability to study the interior structure of disordered systems is the SAS approach's greatest strength, and typically employing this method is a distinctive way to gain immediate structural information on structures with density inhomogeneities that are randomly created on such enormous sizes. Due to its extensive range of explored topics, well-developed experimental and theoretical techniques, and stand-alone branch of the structural analysis of the matter, the SAS technique has become a well-established methodology. International conferences on SAS research have been established every three years in response to these conditions [12].

Small-angle Neutron Scattering

A method known as small-angle neutron scattering (SANS) makes it possible to study materials with length scales ranging from nanometers to micrometers. To investigate the sizes and forms of particles dispersed in a homogeneous liquid, SANS is very useful. This requires scattering one neutron beam from the material and determining the intensity of the scattered neutrons as a function of the scattering angle [12].

Small-angle X-ray Scattering

Small-angle X-ray scattering (SAXS) is an experimental technique that has gained popularity in the biological world. Several notable advances have occurred during the last few decades, most notably in data collection and processing. Through defining typical particle sizes and shapes, SAXS provides valuable structural analysis and physical data for particle systems spanning from 1-100 nm and beyond. In a SAXS experiment, material is exposed to a certain wavelength of X-rays, which elastically scatter between 0 and 5 degrees to produce an intensity distribution that is averaged across space. For samples containing solid, liquid, or even gaseous components, in situ, static, or dynamic studies can be done. These components' structures might be crystalline, sporadic in orientation, or somewhat organized. Depending on the studied structure, it may be possible to determine the kind and dimensions of the lattice, the size of the pores, the inner surface's makeup, the surface-to-volume ratio, and other factors [23].

Nuclear Reaction Analysis

Materials scientists utilize nuclear reaction analysis (NRA) to evaluate the concentration versus depth distributions of specific target chemical elements in a solid thin film. The NRA's fundamental viewpoint is as follows: In the range of energies we utilize, an incoming ion beam enters a thin layer at the surface and loses energy; interactions between the beam and atomic electron shells are mostly responsible for this halting process (electronic stopping). The arriving particle may react nuclearly with film atoms at a specific depth. The energy of the entering ions determines the energy of the reaction byproducts.

It is possible to determine the depth distribution of the atoms interacting with the incoming ions using the energy distribution of the reaction products. Nuclear processes can be used in the analysis as a substitute for the current methods, which are ineffective in detecting light elements. Moreover, NRA is a very sensitive technology capable of detecting concentrations as low as a few tens of ppm [12].

Raman Spectroscopy

The Raman effect is responsible for the operation of the Raman spectroscopy technique. Suppose the molecules in the tissue are excited by the incoming light. In that case, the tissue will reflect the light of a different wavelength (with a wavelength between 750 and 850 nm). It is possible to detect the chemical composition of atherosclerotic plaques by analyzing the reflected light's wavelength, which is unique to individual chemical components. Using flexible optical fibers, excitation light is delivered to the coronary arteries. Collecting the emitted light allows for differentiation between healthy and diseased tissue. An atherosclerotic plaque's influence on atorvastatin and amlodipine was studied using Raman spectroscopy. An atherosclerotic lesion can be slowed by using atorvastatin and amlodipine.

By combining Raman spectroscopy with other diagnostic methods, we may better assess atheromatous plaque and identify vulnerable plaques. This method's main downsides are a shallow penetration depth and minimal blood absorption. Furthermore, Raman spectroscopy takes a long time to acquire and provides no information on the vessel's geometric shape [23].

X-ray Diffraction (XRD)

Due to their wavelengths (0.2 to 10nm), XRD may be used to study materials because the interatomic spacing of crystals can be analyzed. The average spacing between layers and rows is determined using this method. A single crystal or grain can be oriented using XRD, and its size and shape can be determined using the technique.

An elastic and coherent scattering technique is used to scatter collimated X-ray beams from the periodic lattice using a collimated beam of X-rays. The atomic structure of each crystalline substance is unique. Bragg diffraction peaks are created when the dispersed X-ray beam experiences constructive and destructive interference.

X-rays are produced when charged particles are propelled toward an anode at high speeds. In order to make an X-ray beam, a heated filament must first generate an electron beam, which then has to be collimated and accelerated inside a vacuum tube. This flow of electrons will eventually reach its target, the anode. A high-vacuum container houses the anode to prevent air particles from coming into contact with electrons or X-ray photons being generated. The atomic weight of the components in the material affects the X-rays that the substance absorbs.

The detector picks up x-rays, which are then processed electrically or by a microprocessor. With slight adjustments in angle, one can create their own custom-made "spectra" to study. Additionally, XRD can provide additional information about crystallite size. Particles with an anisotropic or non-uniform size distribution are more difficult to characterize accurately using XRD alone.

A complementary technique such as TEM is therefore recommended in these situations. Furthermore, XRD does not allow for the identification of individual particles. The average particle size of a substance, rather than the equation previously stated, determines particle size. On the other hand, Scherrer's equation fails to consider the possibility of peak widening due to internal particle strain and defects. Only the smallest crystallites can be measured with XRD. In other words, it is not an absolute measurement. XRD patterns may identify a mixture's elemental ratio, crystallinity, and deviation of a single component from its ideal composition and/or structure. Mechanistic and Kinetic investigations have recently created in-situ XRD as a characterization technique to monitor the response in real-time [24].

Extended X-ray Absorption Fine Structure (EXAFS) and X-ray Absorption Near-Edge Structure (XANES) are two types of X-ray absorption spectroscopy (XAS). Energy-dependent absorption coefficients are measured using XAS. Absorption edges for each element correspond to differing electron binding energies, resulting in XAS element selectivity. EXAFS is an extremely sensitive method for determining the chemical state of species, even at low concentrations. XAS spectra can only be obtained with the help of synchrotrons. As a result, it is not a widely used or easily accessible method. Electron excitation of an inner shell electron is studied using XANES to determine the density of electronic states that are empty or partially filled. When Mg_2Ge and $GeCl_4$ combine, germanium NPs are formed. Pugsley *et al.* studied the kinetics and mechanism of this reaction using in situ XAS. GeO_2 and Ge NPs are formed in EXAFS and TEM investigations. The first-neighbor Ge-Ge distance calculated using the EXAFS was 2.45, which agreed with the XRD results. Researchers used in situ EXAFS to study the structural changes around germanium atoms within GeO_2 nano powder. Germanium dioxide underwent a full transition in Sulphur, and the researchers observed that GeS_2 was formed. Requejo and colleagues studied alkyl thiol-capped Pd NPs with sulfur-palladium interactions.

Chemical and electronic investigations of the NPs atomic structure and electrical characteristics revealed that the capping thiol molecules were responsible for Pd clusters' surface and bulk sulfidation [1].

Energy Dispersive X-ray (EDX)

The elemental composition of a sample, as well as its chemical makeup, may be determined with the use of energy-dispersive X-ray spectroscopy. The research was conducted by observing how an X-ray source and a material sample interacted. Understanding the atomic structure of each element and the distinctive X-ray spectra it produces is the driving force behind its identification. High-energy charged particles such as electrons or protons are focused on the subject under investigation to produce X-rays of a certain wavelength. During a sample's rest state, electrons (or electron shells) exist in distinct energy levels or shells that bind to the nucleus. For electrons in inner shells, the incoming beam has the potential to excite them, causing them to be ejected and leaving an electron hole in their wake.

A higher-energy electron from the outer shell enters the vacancy and completes the circuit. An X-ray may be generated due to a large energy difference between the higher and lower energy shell. The quantity and energy of X-rays released by a specimen can be determined by an energy-dispersive spectrometer.

These observations are feasible because the X-ray energy may be linked to variations in the energies of the two shells and their atomic structure, which enables the determination of the elemental composition [25].

Cathodoluminescence

An optical and electromagnetic phenomenon known as cathodoluminescence occurs when electrons strike a luminescent substance, such as a phosphor, releasing photons that may have visible spectrum wavelengths. One well-known example is the production of light from an electron beam scanning the inner surface of a phosphor-coated television screen. Cathodoluminescence reverses the photoelectric activity, which produces electron emission when exposed to photons. Cathodoluminescence is a technique used in scanning or scanning transmission electron microscopes to investigate the composition, optical and electrical characteristics, morphology, microstructure, and chemistry at the micro- and sub-nanoscale. A cathodoluminescence (CL) microscope combines the techniques of light and electron microscopes. Its purpose is to investigate the luminescence properties of thin, polished solid sections exposed to an electron beam. CL microscopy may be used to assess constructions built of well-known materials but with complex combinations in addition to analyzing the material composition. In this situation, the CL intensity might be used to compute the local density of states (LDOS) of a nanostructured photonic medium, with the number of accessible photonic states directly related to the CL intensity. For nanoscale materials with high LDOS changes, such as photonic crystals or complex topologies, this is essential [12].

Nuclear Magnetic Resonance Spectroscopy

A subfield of spectroscopy called nuclear magnetic resonance (NMR) studies the phenomena in which many atomic nuclei retain their magnetic moments and angular momentum in an external magnetic field. The foundation of the NMR phenomena is the hypothesis that the magnetic characteristics of atom nuclei may be used to infer chemical information. The feature of nuclear spin is found in nuclei with odd atomic weights or numbers. They are composed of 1H and 13C. (but not 12C).

The spins of nuclei differ sufficiently from one another for NMR tests to be specific to a single isotope of a given element. As the NMR behavior of the 1H and 13C nuclei gives important information that may be used to infer the structure of organic compounds, organic chemists have used this knowledge. Consider applying the proper frequency of radio-frequency radiation to a sample placed in a magnetic field.

The sample's nuclei would be capable of absorbing the energy. The frequency of radiation needed for energy absorption is influenced by two variables. It first and foremost defines the type of nucleus (1H or 13C). Second, the frequency is affected by the chemical surroundings of the nucleus. A spin-containing nucleus can align with (+) or against an external magnetic field (-). Several nuclei have been discovered to spin on an axis. This spin is connected to an electric charge flow since the nuclei are positively charged. Spin-containing nuclei have a magnetic moment similar to the magnet of a compass needle because circulating charges produce magnetic fields. Similar to how compass needles do in the earth's magnetic field, nuclei tend to shift to a favored orientation when exposed to an external magnetic field. There are more prevalent forms of other uncommon orientations. The quantum principles that the nuclei obey allow some nuclei with a spin quantum number of 1/2 to have precisely two orientations. If the appropriate frequency of radiation is absorbed, transitions between the two energy levels are conceivable [26]. Currently, NMR has evolved into a sophisticated and effective analytical tool with numerous applications in medicine, research, and various industries.

Matrix-Assisted   Laser   Desorption/Ionization   Time-of-flight   Mass Spectrometry

MALDITOF-MS, or matrix-assisted laser desorption/ionization time-of-flight mass spectrometry, has evolved as a popular and versatile technology for evaluating macromolecules of biological origin. The time-of-flight mass spectrum analysis that tracks the quick phytovolatilization of a material placed in a UV-absorbing matrix is the underlying notion behind MALDI-TOF-MS. When samples were combined with the suitable matrix material for MALDI-TOF-MS, laser irradiation ionized and desorbed sample molecules as gaseous ions [27].

THERMODYNAMIC CHARACTERIZATION TECHNIQUES

Thermogravimetric Analysis

Thermogravimetric analysis (TGA) is a technique that evaluates the effects of temperature or time on a substance's mass while subjecting a sample specimen to a specified temperature schedule in a predetermined environment. A precision balance supports a sample pan component of a TGA. Throughout the experiment, the pan in the furnace is heated or cooled. The experiment establishes the mass of the sample. The sample environment is kept under control using the purge gas. As the gas passes over the sample and leaves through an exhaust, it may be neutral or reactive. Properties, including enthalpy, thermal capacity, mass changes, and heat expansion coefficient, are provided *via* thermal analysis. In solid-state chemistry, the thermal analysis examines phase transitions, phase diagrams, solid-state reactions, and heat degradation processes.

Differential Thermal Analysis

The chemical makeup of diverse materials is investigated using differential thermal analysis (DTA), which measures a sample's thermal reactivity as it is heated. The approach is predicated on the notion that heat-related interactions and phase transitions occur when a material is heated. DTA is used to determine the temperature differential among the test sample and an adjacent inert substance. The test piece and inert material both have thermocouples fitted, which enables any temperature variations brought on by the heating cycle to be shown as a sequence of peaks on a moving chart. To identify a material, DTA curves made from elements or compounds that are recognized and unknown are examined. The characteristics of specific substances or materials have an impact on the quantity of heat generated and the temperature at which these changes take place. The amount of a drug contained in the composite sample will also be represented by the area beneath the peaks on the graph; this amount may be identified by comparing the area of a similar peak to the regions of other standard samples assessed under the same conditions. It is standard practice to identify specific minerals and mineral combinations using the DTA approach.

Evolved Gas Analysis

Using a thermogravimetric analyzer with a mass spectrometer to look at and identify evolved gases can provide this additional crucial information. The analysis of thermal stability (degradation) processes, the detection of moisture/solvent decline from a specimen, such as that experienced during the drying or dehydration of a pharmaceutical, and the analysis of trace volatiles in a sample, such as volatile organic content testing, are all common uses for TGA-MS.

Differential Scanning Calorimetry

Imaging with distinction Calorimetry is one of the most often used methods for determining the thermal characteristics of solids and liquids (DSC). The DSC system's measuring cell (furnace), the crucible, and the reference pan—which is typically empty—are all filled with a sample. Phase transitions and/or a material's specific heat can be calculated using a programmed, controlled temperature. Heat-flow values are calculated using the cell's calibrated heat-flow parameters.

Nano Calorimetry

We can assess the heat capacity of minuscule samples down to the level of individual material monograms using the potent technique known as nano calorimetry. The best technique to introduce a few material monograms into the nano calorimeter is vapor deposition. Stable glass has a higher heat capacity than conventional liquid-cooled glass. Stable glasses frequently have lower heat capacities than ordinary glasses, implying that they have different molecular packing. The transformation of glass into a supercooled liquid may be seen by tracking changes in heat capacity. It changes more slowly than normal due to the steady glass's kinetic stability. Nano calorimetry is crucial for understanding their behavior, identifying the spectrum of prospective, stable glasses, and knowing how to alter their characteristics [28].

Brunauere Emmette Teller

Calculating a material's specific surface area and providing an explanation for the physical adsorption of gas molecules on a solid surface are both possible using the Brunauere Emmette Teller (BET) hypothesis. In order to calculate surface area, the BET theory uses multilayer adsorption and frequently uses non-corrosive gases (nitrogen, argon, and carbon dioxide) as adsorbates. It typically uses the static volumetric technique and has surface area-determining gas-flow technology.

OTHER IMPORTANT TECHNIQUES

Fourier Transform Infrared Spectrum (FTIR)

Infrared light scans the samples in FTIR analysis to identify organic, inorganic, and polymeric components. Changes in the absorption band pattern imply a material composition change. The Fourier transform infrared spectroscopy (FTIR) may recognize and categorize unidentified chemical compounds, locate contaminants and additives inside a substance, detect breakdown and oxidation, and uncover undiscovered chemicals. After passing through the interferometer, the detector receives radiation from the sources.

The A/D converter and amplifier amplify and convert the signal to a digital signal, which is then processed by the amplifier. The computer then applies the Fourier transform to the signal received from the input and output devices. Infrared light has a wavelength of around $10,000–100$ cm^{-1} and is shone through the sample, with part of it being absorbed and the rest passing through. Vibrational or rotational energy can be generated from radiation absorbed by the sample. From 4000 to 400 cm^{-1}, the molecular fingerprint of a sample can be discerned by the spectrum generated at the detector. Because each molecule has its unique fingerprint, FTIR is indispensable for identifying chemicals [29].

Nanoparticle Tracking Analysis

NPA, a method for seeing and analyzing particles in liquids, is related to the Brownian motion velocity and particle size. The liquid's viscosity and temperature are the only factors that affect the pace of movement; refractive index and particle density have no bearing on it. NTA may be used to determine the particle size profile of microscopic particles in suspension ranging from 10 to 1000 nm.

Tilted Laser Microscopy

A microscope may easily be converted into an instrument for measuring static light scattering, and DLS employs scattering in near-field techniques, which is quite useful for quantifying NP dispersion. A microscope, an objective with a high numerical aperture, and laser light are used in this procedure. The collimated beam is positioned at a broad angle about the objective's optical axis [30].

Turbidimetry

The loss of transmitted light intensity caused by the scattering of suspended particles is measured by turbidimetry. A solution-filled cuvette is passed through after light has been passed through a filter to create light with a specified wavelength. The light that travels through the cuvette is captured by a photoelectric cell. After that, the amount of absorbed light is measured. Turbidimetry analyzes the quantity of light that is transmitted through suspended particles and calculates the amount of light that is absorbed to estimate the substance's concentration. The number of particles and their size determine (1) how much light is absorbed and, consequently, the concentration. Biology may utilize turbidimetry to determine how many cells are present in a solution [12].

Field-Flow Fractionation

A flexible separation technique called field-flow fractionation (FFF) depends on how the flow and the distribution of the field behave in an open, thin channel. Flow FFF (F4) separation is becoming increasingly popular for size-sorting and isolating NPs for further analysis or size/spectroscopic characterization utilizing online, uncorrelated detection approaches. The flow-assisted method F4 is a member of the FFF family. For isolating dispersed analytes with sizes ranging from nanometers to micrometers, this technique is excellent.

Size-Exclusion Chromatography

The separation technology known as size-exclusion chromatography is based on the molecular size of the components (SEC). The sample molecules move across a bed of porous particles and are differentially excluded from the packing material's pores. High biomolecular activity retention is made possible by the material's mild, nonabsorptive contact, SEC's key characteristic. Several different complicated NP samples have been sized using SEC.

Hydrophobic Interaction Chromatography

The most recent addition to the different chromatography modes as of this writing is hydrophobic interaction chromatography (HIC). Most proteins contain hydrophobic regions or patches on their surface, as do hydrophilic molecules (such as DNA and carbohydrates) to a lesser extent. These patches do not dissolve well energetically and leave behind hydrophobic holes in the mobile water phase. The hydrophobic regions of proteins bind to the hydrophobic regions of solid supports when the hydrophobic effect is stimulated (by adding lyotropic salts). As a result of the decreased number and volume of individual hydrophobic cavities, this is thermodynamically advantageous. Lowering the quantities of lyotropic salts lessens hydrophobic interactions and causes desorption from the solid support. Proteins attach to HIC in a special way where they elute at low salt concentrations and bind at high salt concentrations. A reverse salt gradient results from this, which makes it clear that HIC is being used right away.

CONCLUSION

This chapter presents several techniques often used for characterizing NPs, as well as the fundamentals of nanomaterials. This chapter discusses reliable methods for characterizing nanomaterials, including optical (imaging), electron probe, photon probe, ion particle probe, and thermodynamic characterization methods. The concise explanation of each technique's benefits and drawbacks serves as a guide for choosing the best methods for describing nanomaterials. Each strategy has benefits and drawbacks. Scientists are working on solving those issues. The regulated synthesis of NPs and their use depend on an understanding of characterization. On the other hand, it is governed by factors like developing better, faster, simpler, and more efficient methods for characterizing materials. The future of nanotechnology holds great potential. It is necessary to integrate multiple approaches in order to understand particles and their characteristics better.

REFERENCES

[1]   S. Mourdikoudis, R.M. Pallares, and N.T.K. Thanh, "Characterization techniques for nanoparticles: comparison and complementarity upon studying nanoparticle properties", *Nanoscale,* vol. 10, no. 27, pp. 12871-12934, 2018.
        [http://dx.doi.org/10.1039/C8NR02278J] [PMID: 29926865]

[2]   J. A. Drazba, "Introduction to confocal microscopy", *Microsc. Microanal.,* vol. 12, pp. 1756-1757, 2006.
        [http://dx.doi.org/10.1017/S1431927606068280]

[3]   B. Hecht, B. Sick, U.P. Wild, V. Deckert, R. Zenobi, O.J.F. Martin, and D.W. Pohl, "Scanning near-field optical microscopy with aperture probes: Fundamentals and applications", *J. Chem. Phys.,* vol. 112, no. 18, pp. 7761-7774, 2000.
        [http://dx.doi.org/10.1063/1.481382]

[4]   W. Denk, J.H. Strickler, and W.W. Webb, "Two-photon laser scanning fluorescence microscopy", *Science,* vol. 248, no. 4951, pp. 73-76, 1990.
        [http://dx.doi.org/10.1126/science.2321027] [PMID: 2321027]

[5]   R. Xu, "Light scattering: A review of particle characterization applications", *Particuology,* vol. 18, pp. 11-21, 2015.
        [http://dx.doi.org/10.1016/j.partic.2014.05.002]

[6]   S.K. Brar, and M. Verma, "Measurement of nanoparticles by light-scattering techniques", *Trends Analyt. Chem.,* vol. 30, no. 1, pp. 4-17, 2011.
        [http://dx.doi.org/10.1016/j.trac.2010.08.008]

[7]   D. Honig, and D. Mobius, "Direct visuaiizatlon of monolayers at the air-water interface by brewster angle microscopy", *J. Phys. Chem.,* no. 2, pp. 4590-4592, 1991.
        [http://dx.doi.org/10.1021/j100165a003]

[8] K.J. Stine, "Brewster Angle Microscopy", *Supramol. Chem.,* 2012.
[http://dx.doi.org/10.1002/9780470661345.smc040]

[9] T. Khare, U. Oak, V. Shriram, S.K. Verma, and V. Kumar, *Biologically synthesized nanomaterials and their antimicrobial potentials.* vol. 87. 1ˢᵗ ed. Elsevier B.V., 2019.
[http://dx.doi.org/10.1016/bs.coac.2019.09.002]

[10] M. Farré, and D. Barceló, "Chapter 2 - Introduction to the Analysis and Risk of Nanomaterials in Environmental and Food Samples", In: *Comprehensive Analytical Chemistry* vol. 59. Elsevier, 2012, pp. 1-32.
[http://dx.doi.org/10.1016/B978-0-444-56328-6.00001-3]

[11] K. Akhtar, S. A. Khan, S. B. Khan, and A. M. Asiri, "Scanning electron microscopy: Principle and applications in nanomaterials characterization", In: *Handbook of Materials Characterization.* Springer International Publishing, 2018.
[http://dx.doi.org/10.1007/978-3-319-92955-2_4]

[12] C. Jose Chirayil, J. Abraham, R. Kumar Mishra, S.C. George, and S. Thomas, "Instrumental Techniques for the Characterization of Nanoparticles", In: *Thermal and Rheological Measurement Techniques for Nanomaterials Characterization.,* S. Thomas, R. Thomas, A.K. Zachariah, R.K. Mishra, Eds., Elsevier, 2017, pp. 1-36.
[http://dx.doi.org/10.1016/B978-0-323-46139-9.00001-3]

[13] A. Laberrigue, "Experimental high-resolution electron microscopy by J. C. H. Spence", *Acta Crystallogr. A,* vol. 39, no. 3, pp. 503-504, 1983.
[http://dx.doi.org/10.1107/S0108767383000963]

[14] K.E. MacArthur, "The use of annular dark-field scanning transmission electron microscopy for quantitative characterisation", *Johnson Matthey Technology Review,* vol. 60, no. 2, pp. 117-131, 2016.
[http://dx.doi.org/10.1595/205651316X691186]

[15] P.K. Ghosh, *Introduction to photoelectron spectroscopy / Pradip K. Ghosh.* Wiley: New York, 1983.

[16] I. Pastoriza-Santos, J. Pérez-Juste, and L.M. Liz-Marzán, "Silica-coating and hydrophobation of CTAB-stabilized gold nanorods", *Chem. Mater.,* vol. 18, no. 10, pp. 2465-2467, 2006.
[http://dx.doi.org/10.1021/cm060293g]

[17] R. Begum, Z.H. Farooqi, K. Naseem, F. Ali, M. Batool, J. Xiao, and A. Irfan, "Applications of UV/Vis Spectroscopy in Characterization and Catalytic Activity of Noble Metal Nanoparticles Fabricated in Responsive Polymer Microgels: A Review", *Crit. Rev. Anal. Chem.,* vol. 48, no. 6, pp. 503-516, 2018.
[http://dx.doi.org/10.1080/10408347.2018.1451299] [PMID: 29601210]

[18] R.E. Hummel, and T. Dubroca, *Differential Reflectance Spectroscopy in Analysis of Surfaces.* Encycl. Anal. Chem, 2000, pp. 1-25.
[http://dx.doi.org/10.1002/9780470027318.a2504.pub2]

[19] I. Khan, K. Saeed, and I. Khan, "Nanoparticles: Properties, applications and toxicities", *Arab. J. Chem.,* vol. 12, no. 7, pp. 908-931, 2019.
[http://dx.doi.org/10.1016/j.arabjc.2017.05.011]

[20] S. Greenfield, "Inductively coupled plasmas in atomic fluorescence spectrometry. A review", *J. Anal. At. Spectrom.,* vol. 9, no. 5, pp. 565-592, 1994.
[http://dx.doi.org/10.1039/ja9940900565]

[21] J.R. Lakowicz, *Principles of fluorescence spectroscopy, 3ʳᵈ Principles of fluorescence spectroscopy.* 3ʳᵈ ed. Springer: New York, USA, 2006.
[http://dx.doi.org/10.1007/978-0-387-46312-4]

[22] M.H. Herman, "Applications of Rutherford backscattering spectrometry to refractory metal silicide characterization", *J. Vac. Sci. Technol. B,* vol. 2, no. 4, pp. 748-755, 1984.
[http://dx.doi.org/10.1116/1.582873]

[23]   A. Synetos, and D. Tousoulis, "Invasive Imaging Techniques", In: *Coronary Artery Disease* Elsevier Inc., 2017, pp. 359-376.
[http://dx.doi.org/10.1016/B978-0-12-811908-2.00018-0]

[24]   M. Kaliva, and M. Vamvakaki, *Materials Characterization.* vol. 1242. Elsevier Inc., 2010.
[http://dx.doi.org/10.1002/9780470172919.ch8]

[25]   J. Abraham, B. Jose, A. Jose, and S. Thomas, "Chapter 2 - Characterization of green nanoparticles from plants", In: *Phytonanotechnology Challenges and Prospects Micro and Nano Technologies* Elsevier Inc., 2020, pp. 21-39.
[http://dx.doi.org/10.1016/B978-0-12-822348-2.00002-4]

[26]   H. Günther, *NMR Spectroscopy - Basic Principles.* Concepts, and Applications in Chemistry, 2013.

[27]   L.F. Marvin, M.A. Roberts, and L.B. Fay, "Matrix-assisted laser desorption/ionization time-of-flight mass spectrometry in clinical chemistry", *Clin. Chim. Acta,* vol. 337, no. 1-2, pp. 11-21, 2003.
[http://dx.doi.org/10.1016/j.cccn.2003.08.008] [PMID: 14568176]

[28]   J.H. Perepezko, T.W. Glendenning, and J.Q. Wang, "Nanocalorimetry measurements of metastable states", *Thermochim. Acta,* vol. 603, pp. 24-28, 2015.
[http://dx.doi.org/10.1016/j.tca.2014.06.017]

[29]   D. Titus, E. James Jebaseelan Samuel, and S.M. Roopan, "Nanoparticle characterization techniques", In: *Nanoparticle characterization techniques.* Elsevier Inc., 2019, pp. 303-319.
[http://dx.doi.org/10.1016/B978-0-08-102579-6.00012-5]

[30]   D. Brogioli, D. Salerno, V. Cassina, and F. Mantegazza, "Nanoparticle characterization by using tilted laser microscopy: back scattering measurement in near field", *Opt. Express,* vol. 17, no. 18, pp. 15431-15448, 2009.
[http://dx.doi.org/10.1364/OE.17.015431] [PMID: 19724541]

CHAPTER 3

How Nanoparticles Enter the Human Body and their Effects

Abstract: The new scientific innovation of engineering nanoparticles (NPs) at the atomic scale (diameter<100nm) has led to numerous novel and useful wide applications in electronics, chemicals, environmental protection, medical imaging, disease diagnoses, drug delivery, cancer treatment, gene therapy, *etc*. The manufacturers and consumers of nanoparticle-related industrial products, however, are likely to be exposed to these engineered nanomaterials, which have various physical and chemical properties at levels far beyond ambient concentrations. These nanosized particles are likely to increase unnecessary infinite toxicological effects on animals and the environment, although their toxicological effects associated with human exposure are still unknown. These ultrafine particles can enter the body through skin pores, debilitated tissues, injection, olfactory, respiratory, and intestinal tracts. These uptake routes of NPs may be intentional or unintentional. Their entry may lead to various diversified adverse biological effects. Until a clearer picture emerges, the limited data available suggest that caution must be exercised when potential exposures to NPs are encountered. Some methods have been used to determine the portal routes of nanoscale materials on experimental animals. They include pharyngeal instillation, injection, inhalation, cell culture lines and gavage exposures.

Keywords: Intestinal tracts, NPs uptake, Respiratory system, Toxicological effects.

INTRODUCTION

"Nanotechnology" encompasses manipulating matter on a near-atomic scale to produce new structures, materials, and devices. It builds nanoparticles (NPs) whose diameter is below 100 nm by manipulating matter at the atomic level. According to Stern and McNeil, NPs can be engineered or incidental, depending on their origin. Engineered NPs such as quantum dots, dendrimers, carbon nanotubes, and fullerene, which have diameters<100 nm, can be compared to the sizes of living things. Also, NPs like diesel particles are generated incidentally, while living things like viruses are natural living cells with diameters<100 nm. Technology can be applied to biological systems, living organisms, or derivatives thereof, to make or modify products or processes for specific use at the nanoscale

Seyed Morteza Naghib and Hamid Reza Garshasbi

levels. It, therefore, encompasses a wider range and history of procedures with useful industrial and biological processes in modifying the needs of humanity at the nanoscale level. Some studies have also shown that microorganisms can as well be used as potential developers of NPs. With the development of these new approaches and techniques, nanotechnological industries are acquiring new horizons enabling them to improve the quality of products and life with uncertain health safety issues. NPs can enter the environment and animals' systems through different pathways. For instance, it could be through effluent, spillage, consumer products, and disposal. The intake is usually tolerated by the organism's system, but when a certain range is exceeded, it causes toxic effects and even deaths. Since NPs can cause risks to the environment and human health, therefore, research must be undertaken to understand and anticipate such risks through risk assessment and risk management. However, given the limited amount of information about the health risks of NPs, it is prudent to take measures to minimize workers' exposure to the environment.

REGULATION OF NANOMATERIALS RISK ASSESSMENT

Regulation EC No. 1907/2006 concerning the Registration, Evaluation, Authorization, and Restriction of Chemicals (REACH) governs all chemicals and their usage in goods; no special regulation exists in the European Union. Chemicals such as SiO_2 and TiO_2 are extensively used to create coatings and composites, among other things. Even though they are not explicitly addressed, NMs are part of the REACH framework. This is due to including all chemical compounds, regardless of their form or arrangement. REACH mandated the creation by the European Chemicals Agency (ECHA) of guidelines for the disclosure of information and the conduct of safety assessments. Toxicological and toxicokinetic data and appropriate safety evaluations are required for substances that come under the EC's guidelines for classifying NMs.

Also included is an evaluation of nano-specific occupational exposure, with suggestions for protective gear if other measures are ineffective. mg/cm^2, cm^2/m^3, and particle number/cm^3 are proposed as appropriate dosage metrics for inhalation exposure to NM. For fibers, this is especially true. Fiber less than 3 m in length and less than 20 nm in diameter is considered hazardous since it does not biodegrade in the lungs. Carbon nanotubes, silicon carbide, and fluoro-edenite have been associated with cancer development when inhaled through the parietal pleura. NMs used in biocidal goods, whether as active or inactive ingredients, must be approved by an independent risk assessment focused on nanotechnology. The intended application areas must be specified, such as antimicrobial product provision. Antimicrobials used in treating patients have developed resistance to the bactericidal nanosilver (nano-Ag), which has prompted the advice that they must not be used in consumer products due to evidence of their accumulation in humans and the development of bacterial cross-resistance.

In case of cosmetics, this is a specifically controlled product. A cosmetic product must notify the European Commission (EC) about the toxicological profile of the NM and necessary safety data before it may be sold in Europe. According to the European Commission's suggestion, NM is defined in a way that excludes the idea of a nanoscale proportion, which is important. A rigorous safety evaluation is required for all NM-based UV filters, colorants, and preservatives before using them. To evaluate and assess the safety of cosmetic compounds, the Scientific Committee on Consumer Safety's Notes of Guidance should be used. The special characteristics of NMs must also be considered. As a UV filter in sunscreens, nano-TiO_2 in concentrations up to 25% was found to have no dermal absorption and consequently no adverse effects on people, whether applied to healthy or sunburned skin. Inhalable powders and sprays are exempted from this rule. We still need to improve the risk evaluation for non-medical devices. Respirable particles' toxicokinetic characteristics must be taken into account. As long as specific requirements are met and a proper risk assessment is carried out, the European Commission has permission to use nano-TiO_2 in cosmetic UV filters. Cosmetic sprays that emit aerosols that can be inhaled constitute a risk. European Union member states and the European Commission approve NMs in agriculture, food, and feed. For food and feed chain risk assessments, EFSA's guideline document for nanotechnology and nanoscience is used. Animal feed, food additives, and packaging materials are just a few of the many uses for which these substances are evaluated in this manner. There are five stages in which an engineered nanomaterial (ENM) needs to be evaluated: (1) as a generated material (pristine state); (2) in food/feed products; (3) in food or feed matrixes; (4) in media used for toxicity testing; and (5) in bodily fluids of humans and animals. Based on the persistence and ingestion of particles, six scenarios were developed to measure exposure in Table **1**.

There is now a revision of the EFSA guidance. When it comes to food and feed safety assessments, there are many technological hurdles that EFSA must overcome. Nano-encapsulates and Ag were the most often used nanomaterials in agriculture, feed, and food. The importance of food additives and touch materials was emphasized.

When conducting risk assessments, it is necessary to provide a comprehensive report of physicochemical parameters and their analytical methodologies, depending on the NM and its measuring environment. Applied procedures must be proven appropriate for the task at hand and capable of producing repeatable results. Even so, this area of analytical chemistry is still relatively new. As a result, it is unsurprising that the analytical issues involved with applying the EC guideline for the classification of NMs hinder food labeling rules, which were imposed for transparency. Regulatory bodies will expect dependable processes for routine testing in order to meet the criteria of the applicable NM. Overdosage may lead to findings unrelated to the material's toxicity, said EFSA Nano Network members, but rather to the large doses of NMs used in food safety research.

Table 1. Agricultural, feed, and food nanomaterials risk assessment case studies.

Case	Scenario Description
1. Preparations/formulations, as advertised, do not include any engineered NPs.	Evidence showing the ENM has been degraded/solubilized using approved analytical techniques to non-nano form should be used for the particular intended use in accordance with the EFSA Guidance for non-nano forms if persuasive. No longer does this ENM Guidance apply.
2. In the absence of food interaction materials, there will be no migration	If there is enough evidence to show that ENMs do not migrate, the risk assessment might be predicated on the fact that there is no ENM exposure through food and hence no toxicological concern.
3. Before ingestion, designed NPs are transformed into a non-nano form in the food/feed matrix.	EFSA Guidance for non-nano forms should be used for the specific intended purpose. The present ENM Guidance will no longer apply if evidence that the ENM has been changed into a non-nano form in the food/feed matrix is complete.
4. Transformation during digestion	If the possibility of ENM absorption before the dissolution/degradation stage can be ruled out, data from the non-nano form material can be used for hazard identification and characterization. If there is convincing evidence that ENM absorption does not occur, *in vitro* genotoxicity study, *in vivo* local effects, and/or other acceptable *in vivo* tests may be sufficient. ENM dissolved in water is predicted to have a similar systemic toxicity profile as its soluble counterpart. If this can be demonstrated, then no more testing of ENM is required. When non-nano form data are not available, EFSA Guidance for the intended application calls for testing in the non-nano form.
5. Information on a non-nano form is available	Using the information on ADME, toxicity, and genotoxicity gleaned from ENM testing on the chemical's non-nano form, the non-nano form should be put to the same rigors as the ENM-derived version. If there are substantial differences between the ADME and toxicity data from both forms, it is important to note them. If the disparities suggest a larger risk, more toxicity testing on the ENM will be required, in addition to the ADME, 90-day, and genotoxicity studies. Scientific justification is required for deferring more testing if the differences indicate a lower risk.
6. No information on a non-nano form is available	A non-nano version of ENM may not be available for toxicity testing; hence EFSA's advice for intended use should be followed with the revisions below to consider the nano properties of ENM in cases where ENM stays in the food/feed matrix and gastrointestinal fluids. The ENM toxicity testing technique takes nano features into account while identifying and characterizing potential hazards.

It is important to use dosages that are physiologically realistic for testing purposes. This means that risk evaluations for novel foods, as well as those for food contact materials and additives, must be carried out on a case-by-case basis. Particle migration from nanocomposites, commonly used in food packaging and textiles, is a special issue. There are no nano-specific rules that apply to cleaning agents or detergents, except for the REACH application of their ingredients, which textiles are not. Suppose loose NPs are released into the air or touch human skin or lungs.

In that case, they are regarded as a possible health hazard because there is still much uncertainty regarding the dangers that particulate matter can have in the immediate area [1].

NPs Circulation inside the Body and Interact with Biomolecules

Inhalation, ingestion, and skin contact are all routes by which they can enter the human body due to their minuscule size. Different internalization pathways allow NPs to enter cells, accumulating in specific tissues before ejection. In the extracellular fluid, they are connected to biomolecules that allow them to enter cells.

Cellular Internalization

NP-based nanomedicines cannot diffuse into cells since cell membranes are impermeable. This means that they must first breach the membrane by a variety of mechanisms, including micropinocytosis, Clathrin endocytosis, caveolae-dependent endocytosis, or direct penetration. Non-targeted internalization mechanisms can also cause toxic effects. Internalization is a possible approach when biocompatibility is desired. The internalization process is directly affected by the NP size, as expected. NPs in the 10–100 nm range have a higher cellular absorption, while smaller ones have a higher energy cost for the cells. For example, Dendritic cells in macrophages can take in NPs larger than 100 nm, allowing for more precise targeting.

Zeta potential and endocytosis/exocytosis pathways are linked in several studies. In NP interactions with biomolecules and cells, van der Waals and electrostatic forces are critical. Furthermore, it is vital to highlight that the NPs' capacity to be absorbed into cells is directly tied to the NPs' surface chemistry. This means that specialized cell-to-cell interactions, rather than non-specific ones, may be optimized.

The internalization potential of NPs coated with antibodies is as much as four to eight times larger than that of NPs that are positively or negatively charged but do not have an affinity for target cells. There is always a possibility of non-specific interactions *via* chemical moieties that influence target affinities when utilizing antibodies for targeted distribution. The shape of the NP affects protein adsorption, which affects cellular uptake. Cellular uptake of non-geometrically symmetric and amorphous nanoconjugates is significantly reduced. To further prevent non-targeted cell internalization, a form can be used, according to a large number of authors.

Tumor Accumulation

NPs are more likely to accumulate in tumor tissues than in healthy ones [2]. There are two main reasons for this: greater permeability (compared to normal vessels) of the vasculature surrounding the tumoral tissue and the tumors' poor lymphatic outflow. Enhanced permeation and retention describe this phenomenon (EPR). In tumors, the cells and extracellular matrix are packed together densely. Because of this, tumor spread and accumulation are heavily influenced by the NP's size. Nanoparticle (NP) features, such as size and surface chemistry, may influence cancer cell accumulation. In general, there is an inverse relationship between diffusion and NP size [3]. They can easily penetrate tumor tissue and spread across healthy tissue because of their tiny size. The larger the NPs, the more difficult it is to distinguish between healthy and diseased tissues, so they are often utilized as imaging agents to aid tissue differentiation. The ability of biomolecules to opsonize and be cleared from the NP surface is linked to their adsorption onto the NP surface, since blood concentration and time are intertwined.

Elimination

The renal and hepatobiliary systems are the most common ways through which NPs are eliminated from the human body. It is necessary to secure clinical approval in a reasonable amount of time. That is why quick clearance and long-term bodily maintenance cannot coexist in the case of drug-conjugated NPs. As expected, removal is affected by surface chemistry, shape, and NP size. When clearing NPs, the surface chemistry is critical, even for the smallest NPs, and PEG coating enhances hepatobiliary clearance. The NP size is also a consideration. A physical barrier in the kidney, the hydrodynamic NP size, significantly impacts renal clearance.

Nanoparticle Interactions

Biomolecules surround nucleic acids in biological systems. However, the number varies depending on the properties of the biological environment. The NP can foster different kinds of interactions. See Fig. (**1**) for examples of nanomedicines implanted in human blood, cells, and culture medium [4]. Additional design and development issues arise because of the enormous range of microenvironments. There are a variety of ways that ionized nanoparticle aggregates or interact with other nanomaterials in diverse mediums based on pH, ionic strength or oxygen levels, and organic matter. Different media contain NPs in a variety of shapes and stages. Nanomaterials' instability and immuno-biocompatibility may cause a diverse morphology [5].

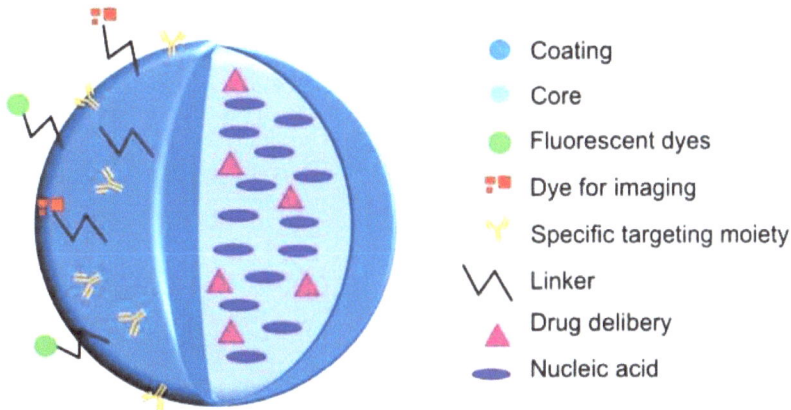

Fig. (1). Schematic representation of multi-functional NPs [6].

It is possible that the NPs composition and kind of cell influenced cell absorption and may cause toxicity. In nanotechnology, agglomeration and agglomeration effects are often misunderstood. An aggregation suggests a strong merging or fusion between two objects. Particles are drawn together by van der Waals forces, which are stronger than the electrostatic forces induced by the nanostructure surface. While this may be true for a certain pattern or size in distribution. Tin oxide NPs' optical characteristics were examined by Pellegrino F. *et al.* Agglomeration and aggregation may lead to inaccurate estimations of photoactivity, according to their study. NP agglomerates were generated utilizing a bottom-up approach in an aqueous media by Zook M.J. *et al.* [7] Silver NPs agglomerate because they wanted to see how. Hemolytic activity was shown using this approach. In nanoparticle interactions, the compatibility of the nanomaterials, the distance between them, and their shape all play a role. NP assemblies' major interaction drivers must also be understood. When exposed to UV light or molecular force, iron oxide NPs containing azobenzene-terminated catechol ligands self-assemble due to magnetic interactions [8]. Material complementarity and the relevance of forces utilized in such an interaction are underlined in a case study by Pileni and associates.

With no Van der Waals interactions, they point out that organic ligands like octanoic and dodecanoic acids may make a significant impact on the NPs' performance. Surface functionalization is essential for molecules to interact with surfaces. Fig. (**2**) illustrates this point. When more than one chemical group coexists on the surface, reactive chemical moieties are either homo- or heterobifunctional.

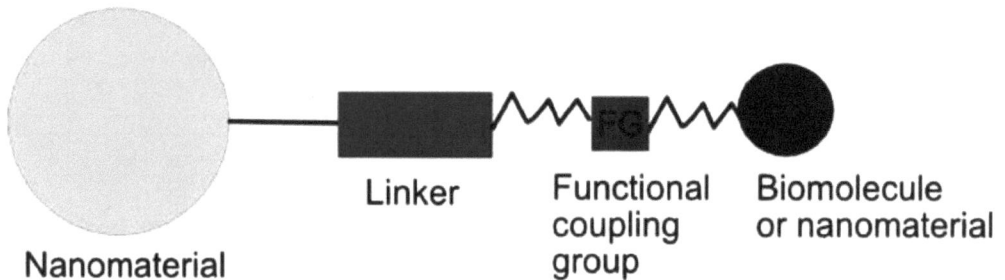

Linker   **Functional   Biomolecule
                    coupling   or nanomaterial
Nanomaterial   group**

Fig. (2). The approach for coupling NPs with biomolecules or other NPs is depicted schematically [6].

It is possible that the surface's composition and structure prevent it from supporting a wide variety of interactions. Thus, the lipid bilayer of circulatory cells is enriched with proteins and polysaccharides that favor one sort of contact mechanism [9]. Another example is the influence of a protein's molecular weight, charge, or stability on the number of binding sites it can have. There are more active centers to interact with and less structural stability in a soft protein layer and other surface-influencing physicochemical variables. A protein's reactivity and adsorption characteristics may be influenced by its hydrophobicity and hydrophilicity surface ratio. Many are smaller than NP in size, which makes it easier for them to be taken in. No matter how high or low their concentration or size, one must always keep in mind that NPs are more than just a collection of particles.

Interaction Mechanisms between NPs and Biomolecules

More than a hundred different biomolecules could interact with the NP's surface directly or indirectly *via* other biomolecules on the NP's surface (Fig. 3). It is important to note that the kind of organism, biological fluid, cells, and so on all have a direct impact on the NP surface, nature, and structure of biomolecules in these coatings.

Proteins and nucleic acids are the macromolecules that interact with NP surfaces the most often, according to the research. Non-specific or specific adsorption results in several protein binding sites. In addition, proteins play a crucial role in the nanomaterials' immuno-biocompatibility. Molecular nano-construction may use nucleic acids' high base pairing selectivity, physicochemical stability, mechanical stiffness, and ease of accessibility [11]. Consider two factors when discussing interactions between human biomolecules and other organisms [12].

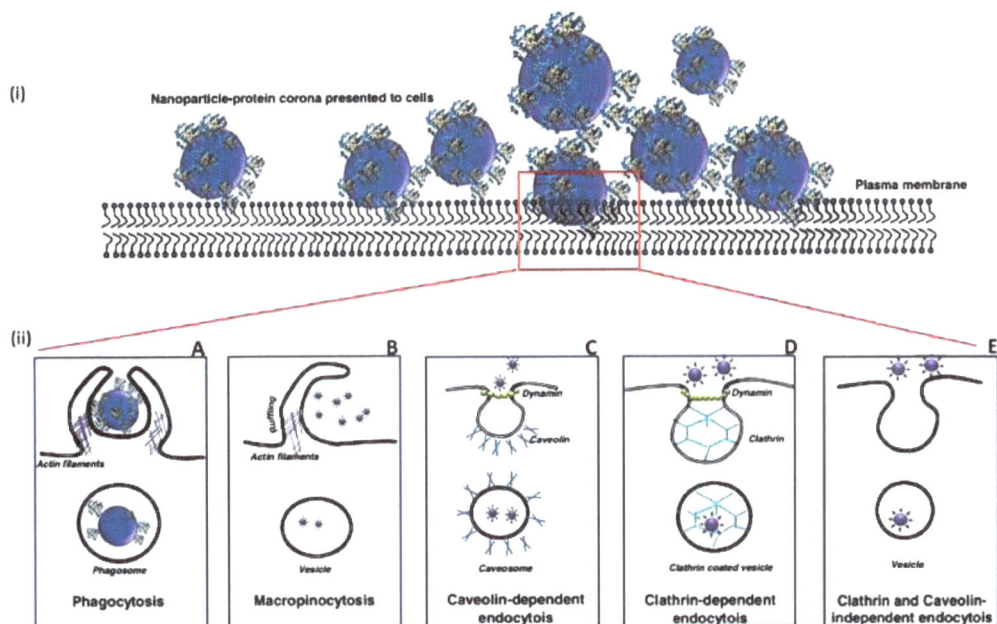

Fig. (3). The interface between NPs and cell membranes. The protein corona serves as a bridge between NPs and cells. (**A**) In case of massive NP-protein complexes, macrophages and neutrophils may phagocytose these NP agglomerates. The phagosome is formed by folding the plasma membrane over the NP complex. (**B**) Additionally, macropinocytosis, which is how cells take in extracellular fluid containing aggregates of NP, may also be used to take them up. Specific receptors involved in the production of NP complexes may also drive endocytosis. (**C**) caveolae plasma membrane indentations consist of cholesterol-binding proteins called caveolins or (**D**) clathrin-coated vesicles. (**E**) Clathrin or caveolae, in addition to these other endocytic processes, may promote NP uptake [10].

In biological systems, several potentially interacting biomolecules surround and saturate the surface of NPs, the first. As a result, only NPs that have undergone certain modifications can interact with biomolecules of interest in the future. In addition, the human body is exposed to NP through many channels. The strength of the interaction will be affected if this is the case. Inhaling NPs, for instance, has a significant impact on the respiratory system (proteins and phospholipids). Various biomolecules have been used to study two methods of immobilization: easy absorption and chemical coupling. Although enzyme adsorption on NPs can disrupt the enzyme's active site, it is advantageous since it uses non-covalent forces to immobilize enzymes on the NPs. Immobilization of biological molecules can be achieved without interfering with their natural structure or affecting their biological function using chemical linkages.

Cells can also communicate through ligand-receptor interaction and chemical conjugation [13]. The streptavidin-biotin-functionalized surface of NPs is an example of the first way to interact with NPs. As a result, it is more resistant to changes in pH, temperature, and denaturants due to its stronger non-covalent interaction. In addition, they are more likely to adhere to cells.

When functional groups (like thiols) are attached to the NP surface, it boosts the NP's ability to bind to cells in the future, reducing toxicity. Although this method has potential biological applications, one drawback is that medication covalently attached to the NP prevents its efficient release, reducing its effectiveness [6].

NPS ROUTES OF ENTRY, EFFECTS ON THE HUMAN BODY, AND TOXICITY

Respiratory Tract Uptake and Clearance

During their manufacturing or use, nanomaterials significantly impact human health due to their toxicity to the skin and lungs. When the NPs inhaled during this procedure are deposited in alveolar areas, they can be observed. Alveolar deposits often contain nanomaterials with a diameter of 10–100 nm. The respiratory system mostly prevents foreign particles from entering the airway epithelium. The respiratory wall consists of a mucus layer and a surfactant film. Tight connections hold ciliated cells of the bronchiolar epithelium in place. Smokers' nonciliated alveolar area is where most of the inhaled NPs end up [14]. While epithelial cells pick up these, alveolar macrophages reach the mucociliary escalator and do not. If the number of NPs inhaled exceeds the macrophage's ability to remove them, the cells are damaged. The NPs' toxicity is limited to the lungs, and their retention and systemic absorption are minimal in the lungs. NPs have been shown to pass the pulmonary barrier, enter the circulation, and demonstrate toxicity, according to some researchers. The size and concentration of NPs determine their capacity to pass the lung barrier [15].

Cellular Interaction with NPs

Endocytosis is the process through which NPs are taken in by cells [16]. Endocytosis is when cells ingest molecules from the environment by encasing them in their cell membranes. In the scientific community, phagocytosis and Pinocytosis are the most commonly used terms to describe this process. The cellular phenomenon of phagocytosis outlines how phagocytes (specialist cells like macrophages) eliminate foreign particles in the blood, such as NPs (nucleic acids) [17]. Particles coated with transferrin enter cells through the caveolae pathway, facilitated by endothelial cells in the brain. For permeable NPs, the same has been seen [18]. Pinocytosis is the most often used method for ingesting gold

NPs. Endocytosis through clathrin/caveolin independent mechanisms and clathrin/caveolin dependent mechanisms are all important in this process. Plasma membrane extensions that resemble lamellipodia are produced by the process known as macropinocytosis [19].

NPs greater than 200 nm may infiltrate cells *via* macropinosomes. PEGylated-poly-L-lysine NPs are taken up through macropinocytosis in cells. The endocytic system relies heavily on receptor-mediated processes. Receptor-bound NPs are wrapped around by the cellular membrane, which then pinches off to create vesicles [20]. Proteins like clathrin or caveolae help this process; 100-200nm clathrin coats and 50-80 nm diameter spherical caveolae [21]. There are clathrin-coated pits that can only hold NPs smaller than 100nm [22]. The clathrin-dependent endocytosis of NPs is restricted to ligands attached to receptors. The internalization of NPs was more efficient when the particles were smaller than caveolae.

The size of NPs is an essential element in cellular uptake, according to various research works on targeted medicine delivery into cells. More than five to ten times as many cells took up 20-40 nm NPs as 100 nm. Larger particles cannot be taken up by caveolae because of their size [23].

It is common for virus budding to occur at a certain level of internalized particles [24, 25]. When it comes to endocytosis, it is important to remember that spherical particles should have an optimal radius between 27 and 30 nanometers (nm) [26, 27]. For receptor-mediated endocytosis to work correctly, particles should be as small as possible but not smaller than this. At the cell-particle interface, various factors impact particle endocytosis, including the binding energy and bond elasticity, as well as non-specific attraction and repulsion forces [28, 29]. The endocytic routes for NPs to enter cells are governed by their electrical charge. These paths could be employed to internalize NPs, depending on their size, shape, and composition. Negatively charged NPs cannot utilize the clathrin-mediated endocytosis mechanism. Clathrin-mediated absorption of positively charged NPs occurs quickly [30]. Endocytosed NPs are typically confined to endosomes. Liposomes or modified NPs that include cell-penetrating peptides on their surface can inhibit NPs from entering the endosome [31].

NP uptake is determined by the size, surface charge, and ligands bound to the cell surface. Endocytosis is used in tumor cells to ingest polycaprolactone/PEG/polycaprolactone NPscarrying doxorubicin [32]. Polyethyleneimine and poly amidoamine dendrimers, two NPs having positively charged groups on their surface, have the potential to disrupt the plasma membrane. Nanoholes are formed when the plasma membrane is damaged. NPs have accumulated on the cell surface in addition to forming holes [33]. At times, this form of cell-to-cell NP aggregation can cause cell dysfunction. Smaller NPs puncture cell membranes, while larger NPs form a protective lipid bilayer around them [34]. Modified glycol chitosan NPs have been studied for their cellular absorption process and intracellular destiny. Interestingly, these NPs had a greater cell dispersion than hydrophilic glycol chitosan polymers [35].

Endocytic inhibitors allow glycol chitosan NPs *in vitro* to enter cells by clathrin or caveolae-mediated endocytosis, macropinocytosis, or any combination of these three mechanisms. Lysosomes also include some glycol chitosan NPs, which have been trapped. Lysosomes trap DNA NPs after they have passed through various biological processes, just like poly-L-lysine-PEG [36].

Nervous System Uptake of NPs

Cellular uptake is mainly accomplished through endocytosis. For metallic NPs to enter the cell, they must first interact with the membrane. Endocytic vesicles transport NPs to specified intracellular compartments after membrane invaginations and budding and pinching to generate endocytic vesicles. The substances involved in endocytosis are broken down into numerous categories.

Pinocytosis and phagocytosis are two methods of endocytosis. It is possible to classify pinocytosis into four types based on vesicle size and proteins involved in their formation. Pinocytosis can be classified as clathrin-mediated, caveolae-dependent, clathrin/caveolae-independent, or macropinocytosis [37]. On the other hand, Pinocytosis occurs in a far broader range of cells than phagocytosis. Interestingly, the endocytosis-based uptake of metallic NPs in neurons and glial cells encompasses all types of uptake, even phagocytosis. In astrocyte-rich primary cultures, Ag NPs were taken up by endocytosis. The endocytosis of ZnO NPs by PC12 neuronal cells was needed for NP translocation into interneurons, as demonstrated in this study. Astrocytes also took up the iron NPs (Fe NPs) coated with dimercaptosuccinic acid. It was found by employing transmission electron microscopy that they were gathered into intracellular vesicles. They have a negative charge, according to their zeta potential.

Endocytosis mediated by clathrin results in clathrin-coated vesicles being formed. Nutrients and membrane components are taken in by the body during cellular uptake. Several processes rely on caveolin-dependent endocytosis in addition to endocytosis itself, including signaling and transcytosis. Endocytosis independent of clathrin/caveolae occurs *via* various methods when these proteins are not involved. It is a type of Pinocytosis in which the cytoskeleton is rearranged to cause membrane expansion or ruffles. Regardless of specific receptors, extracellular fluid containing dissolved compounds is collected.

Microglial cells took up Fe NPs mostly by macropinocytosis and clathrin-mediated endocytosis, according to Luther and colleagues. Endocytosis was the predominant cell process in neural stem cells cultured with Ag NPs. Endocytosis was also cited as a factor in the absorption of TiO_2 NPs by glial cells. TiO_2 NP absorption was significantly reduced when the cytoskeleton was chemically immobilized, suggesting that macropinocytosis is the major mechanism. The cell's interior can be reached by phagocytosis and various pinocytosis mechanisms. Immunoglobulins or other blood proteins opsonize NPs, allowing them to be detected by cells capable of detecting them. Engulfment and internalization of NPs, known as "phagosomes," are enabled by this signaling cascade. TiO_2 NPs led to a distinct phenotype of activated microglia cells. Phagocytosis-typical morphological modifications, such as increased size and the development of membrane protrusions, suggested TiO_2 NP uptake. According to recent studies, neurons and glia-like cells may absorb metallic NPs *via* a variety of endocytosis mechanisms. In ALT cells, clathrin- and caveolae-dependent phagocytosis and clathrin-mediated endocytosis were employed to take up TiO_2 NPs and BV-2 microglia, respectively. In BV-2 cells, clathrin-mediated endocytosis was used. In a similar investigation, neuroblastoma N2a cells absorbed fewer Ag NPs than ALT and BV-2 cells.

Phenomena in ALT and BV-2 cells differed greatly regarding endocytosis, with phagocytosis predominating in the former and micropinocytosis predominating in the latter [38].

Nanoparticle Translocation to the Lymphatic Systems

In the GI tract's epithelial lining, enterocytes (cells that absorb nutrients) and goblet cells (secrete mucus) stand out as distinctive cell types. A mucosal layer shields these cells from the outside world, forming a solid barrier. The enterocyte layer of the GALT contains lymphoid follicles, which create the mucosal immune response.

Various animals and people have Peyer's patches, which contain these follicles, depending on their age. Enterocytes, M cells and other cells make up the epithelium of these follicles (FAE), in addition to goblet cells. These sites serve as the initial line of defense for antigens. The FAE and M cells are the greatest places to accept particles. The presence of M-cells, which aid in endocytosis, intracellular transport, and neighboring lymphoid tissue, is one of their distinguishing features. An M-surface cell's apical membrane is the most common site for nanoparticle attachment. From there, they are swiftly ingested and delivered to the lymphocyte.

Hepatic first-pass metabolism and drug loss can be avoided by using GALT absorption as an alternative. Several factors influence NPS absorption through the GALT, including the size and shape of NPS, the charge on the surface, the chemical stability, interactions with gut contents, transit time through the GIT, transport through the mucosa, adhesion to epithelial surfaces, and particulate aggregation when it comes into contact with gut fluids. The course of transit and NPS translocation is determined by the NPS's average diameter, surface charge, and release parameters.

Certain processes, including aggregation, adsorption, and adhesion, may influence the NPS's zeta potential, hydrophilicity, and size. Because of this, the efficacy of these formulations may be compromised by using a specific drug delivery method, reducing their viability. Additionally, NPS's loading capacity is a major consideration. It is directly correlated between loading capacity and bioavailability that higher loading is related to higher bioavailability per ingested particle. Besides these difficulties, the NPS components' biocompatibility and biodegradability are important in determining the formulation's use.

Many studies have shown the benefits of lymphatic targeting. (i) oral delivery of nano-encapsulated GIT labile molecules, (ii) oral delivery of nano-solubilized poorly soluble molecules, (iii) improved bioavailability of poorly absorbed drugs due to increased residence time and surface specificity of NPS, (iv) oral delivery of vaccine antigens to gut-associated lymphoid tissue, (v) translocation of antineoplastic drugs for lymphomas, and (vi) delivery of diagnostics for the lymphatic system. (vii) sustained/controlled drug release, essential for toxic drugs, (viii) reduction of drug-related GI mucosal irritation and avoidance of the hepatic first-pass effect.

NPs Translocation to the Circulatory System

Inhalation or implantation of metallic NPs smaller than 30 nm has been shown in studies on healthy animals to penetrate the circulatory system rapidly. Nonmetallic NPs, smaller than 4 to 200nm, have little or no possibility of going through. A higher capillary permeability is found in patients with respiratory and circulatory diseases, making it easier for metallic or nonmetallic NPs to reach the blood.

Long-term Translocation

NPs can propagate across the respiratory epithelium after being deposited in the lungs. They can remain in the interstitium for years or infiltrate the lymphatic and circulatory systems after they have passed through the respiratory epithelium. Long-term exposure to organs, including the bone marrow, liver, heart, and kidney, is made possible by the circulatory system. Regardless of the rationale, smaller particles are cleared from the lungs quicker than bigger ones, perhaps due to macrophages' lower efficiency in removing small NPs. Since they are more easily absorbed, they can go through the body more quickly.

Short-term Translocation of Metals

The rapid transport of metal NPs from the lungs to the circulatory system and organs has been demonstrated in animal studies. Findings in the pulmonary capillaries included Au NPs ranging in size from 30 to 22 nanometers, as well as Ag NPs ranging from 15 to 30 nanometers. Titanium dioxide NPs of 22 nm diameter have been identified in rats' cardiac connective tissue fibroblasts in animal investigations. At 30 minutes after intravenous injection of gold NPs 30 nm, platelets in the pulmonary capillaries of rats were shown to contain large levels of 30 nm gold NPs.

Short-term Translocation of Nonmetals

Whether or not radiolabeled NPs may be transported quickly from the lungs to other organs is a hot topic. Carbon-based nanomaterials have not been shown to move quickly into the body's circulation in any conclusive way. Radiolabels detach from their tagged NPs. There may not be real mobility of NPs but radiolabel migration throughout the body. Short-lived isotope ^{99m}Tc is used to mark NPs injected or inhaled by individuals and has an atomic diameter of 0.37nm.

The radiolabel may detach from the NPs and follow an alternative path to the intended target. Pertechnetate (^{99m}TcO$_4$) has a slightly greater diameter than the radioactive label in the presence of oxygen. For radiolabeled NPs ranging from 56 to 200 nm in diameter, there is little or no translocation into the circulation in most studies; however, there is substantial translocation for particles ranging from five to ten nm in diameter and twenty to thirty nm in diameter, which is mainly found in the bloodstream. Short-term extra-pulmonary translocation into circulation in healthy individuals is still a hot topic. Nevertheless, it seems that pulmonary inflammation and enhanced microvascular permeability may expedite nanoparticles' translocation into the bloodstream. If NPs can cross the lungs and circulatory system, they may cause more harm to patients with respiratory or blood issues.

NPs Interaction with and uptake by Blood Cells

Blood cells can absorb NPs in distinct ways. The three types of blood cells include red blood cells, white blood cells, and platelet-forming platelets. There are no phagocytic receptors on red blood cells. Therefore their ability to take up NPs is solely dependent on their size. On the other hand, nanoparticle charge or material type is of little significance. Polystyrene particles that are not charged do not alter the formation of blood clots. Platelet uptake and blood clot formation are influenced by the charge of NPs. There is a considerable difference between negatively and positively charged NPs in the production of thrombi and the formation of thrombosis. As a result of their net negative charge, platelets can interact with positively charged particles. A decrease in surface charge and increased aggregation potential result from the negatively charged platelets' interaction with positively charged NPs. Three things are thought to cause blood clots: reduced blood flow, damage to vascular endothelial cells, and altered blood chemistry. Each of these is considered a possible cause of blood clots. However, new research suggests NPs could serve as a blood clot nucleating centers. Installing large 400 nm NPs into the lungs elicited lung inflammation comparable to particles of 60 nm size but did not result in peripheral thrombosis. There is no evidence that thrombosis is caused by larger particles but rather by direct platelet activation, as demonstrated by the inability of larger particles to promote it. Patients with bleeding issues had blood clots that included alien NPs, according to EDS research. Patients with blood disorders who wear vena cava filters, implanted to prevent pulmonary embolism, collected blood clots after six months of use. Gold and silver NPs, cobalt and titanium, antimony, tungsten, nickel, zinc, mercury, barium, iron, and stainless steel are found in the fibrous tissue clots of people with the same blood condition. From nanometers to microns, particle size is the global denominator.

ADVERSE HEALTH EFFECTS OF THE CIRCULATORY SYSTEM

Uptake Thrombosis

Blood clot formation has been linked to nanoparticle migration into the circulatory system. The onset of thrombosis occurs during the first hour of exposure, making this a rapid procedure. In the first hour, NPs of charged polystyrene 60 nm and diesel exhaust particles of 20–50 nm greatly stimulated the development of arterial or venous thromboses. The prothrombotic effects lasted for 24 hours following implantation. Thrombus diameters are correlated with a dose-dependent response to the number of pollutants delivered. Anemia and other blood-related issues would be expected if inhaled NPs were identified in red blood cells in pulmonary capillaries.

Cardiovascular Malfunction

The inhalation of nano- and microparticles has been linked to cardiovascular issues. Inhaled NPs have been shown to cause respiratory problems. However, the link between lung particles and cardiovascular problems is still unclear. A release of cytokines into the blood from the inflammation of the particles' lungs was thought to be responsible for the adverse effects on the heart and circulatory system. This suggests that NPs cardiothoracic effects may be directly linked to their existence in our bodies, as shown by many animal and human studies [12].

Liver, Spleen, and Kidneys: Uptake of NPs

The destiny of NPs delivered *in vivo* is determined by the design of NPs that interact with liver cells. Nanoparticles' physicochemical characteristics, however, have not yet been studied *in vivo* to determine how they affect liver sequestration and cell contact. The aggregation of NPs at the organ level is the primary focus of *in vivo* investigations. Because of this, most research on the structure and location of liver cells in culture does not consider the unique characteristics of each cell in the liver. Parenchymal hepatocytes make up the majority of liver cells. NPs are mostly absorbed by non-parenchymal cells, in any case. Removal of hepatocyte-interacting particles from the body may be accomplished through the hepatobiliary pathway [39].

Smaller than 2 nm, endothelial cells lining the blood vessels constitute a physical barrier to particles. According to the organ or tissue, bigger values have been observed between 50 and 100 nm. The blood-brain barrier refers to the brain's extremely tight endothelial junction. It has been shown that ferritin macromolecules with a diameter of roughly 10 nanometers can enter deep brain tissue when injected into rats' cerebrospinal fluid. For example, the liver's endothelium has pores as small as 100 nanometers, making it easier for larger particles to flow through. Inflammation increases the endothelium's permeability, making it easier for particles to flow through. Patients with orthopedic implants, drug addiction, worn dental prostheses, blood issues, colon cancer, Crohn's disease, ulcerative colitis, and diseases with an unknown etiology have micro-and nanoparticle debris in their organs and blood, according to scanning electron microscopy. Deaths from coal mining had higher levels of particulates in their liver and spleen, according to autopsies. More NPs are found in the organs of sicker workers than in healthy workers' organs. Inhaled NPs are most likely to enter the circulatory system through the lungs, where the body absorbs them. Inhalation of stainless steel welding fumes causes manganese to build up in the blood and liver of rats.

A 30-minute inhalation exposure time in rats with silver NPs measuring 4–10 nm showed that the NPs had already reached the circulatory system. Overnight it can be located in the liver and other organs. After a week, it is removed from these areas. One form of liver clearance is biliary secretion into the small intestine.

Nanoparticle accumulation in the liver and kidney is linked to the wear of dental bridges, according to a case study. The assumption was that the most probable absorption pathway was through the gastrointestinal tract. Particles identified in feces and the biopsies were similar in chemical makeup to those in dental prostheses. The greatest particle size in both organs was 20 micrometers, suggesting that the intestinal mucosa and the liver absorb particles before they are circulated and eliminated by the kidneys. After removing dental bridges, no particles are in the stool [12].

GASTROINTESTINAL TRACT UPTAKE AND CLEARANCE OF NPS

Exposure Sources

NPs in the gastrointestinal tract can come from both internal and external sources. An example of an external source is the excretion of calcium and phosphate from the digestive tract. Affluent countries are anticipated to consume roughly 1012 NPs per person per day in nanoparticle consumption. TiO_2 and mixed silicates make up the bulk of the composition of these materials. Products like salad dressing containing NPs TiO_2 whitener can boost the average daily intake by more than 40-fold. Macrophages store them since they do not decay. In certain cases, mucociliary escalator particles may be eliminated before entering the digestive system. The gastrointestinal tract is also exposed to a tiny percentage of inhaled NPs.

Size and Charge-dependent Uptake

There are many different barriers and exchange mechanisms in the gastrointestinal tract. Macromolecules enter the body through this method. When food enters the small and large intestines, the villi absorb it directly from the epithelium. Since the mid-17th century, nano- and microparticle uptake has been the subject of several studies. Recent years have seen whole issues of scientific journals dedicated to the subject. A particle's size, surface chemistry, and charge, as well as the time of administration and dosage, all affect the amount of absorption in the gastrointestinal system. Particle size affects absorption in the gastrointestinal tract, with smaller particles absorbing more quickly. It has been found that the uptake of polystyrene particles ranges from 6.6 percent for 50 nm particles to 5.8 percent for 100 nm NPs and 0.8 percent for one-meter-long polystyrene NPs.

The lymphatic system and capillaries can be accessed by NPs when they touch the submucosal tissue. The enterocytes of the intestines receive particles that have passed through the mucus layer. An enterocyte is an epithelial cell found in the small and large intestine's superficial layer that aids in absorbing nutrients. Particle size affects the time it takes for NPs to pass the mucus layer in the colon:

smaller particles move more quickly than larger ones NPs with a diameter of 14 nm, 415 nm, and 1000 nm are all unable to breach this barrier.

Diabetes may lead to an increase in the gastrointestinal tract's ability to absorb particles. There was a 100-fold increase in 2μm polystyrene particle uptake by diabetic rats compared to nondiabetic rats. Larger particles can be transported up to 20 μm by inflammation. Due to the gastrointestinal tract's highly charged environment, positively charged latex particles become caught in mucus. Negatively charged latex NPs, on the other hand, diffuse through the mucus and interact with the cells of the epithelium.

Site-specific targeting is now possible thanks to nanoparticle properties like size, surface charge, ligand attachment, and surfactant coatings that can target distinct sections of the gastrointestinal system. It was hypothesized that NPs would not be able to persist in the intestinal system for long periods because of the rapid transit times and the constant renewal of epithelium. Researchers who have studied the effects of ingestion with NPs say the majority is excreted in the feces or urine within 48 hours and 98 percent within 72 hours. According to previous studies, various organs can be affected by NPs, including the brain and the bloodstream. It was shown that oral rats would ingest polystyrene spheres from 50 to 3 nm, found in blood, spleen, and liver. Bone Marrow also included particles. Although there were no particles larger than 100 nanometers in the blood, there were no particles larger than 300 nanometers in the bone marrow.

The heart and lungs were not found to have any particles during the investigation. There was no evidence of iridium absorption despite the presence of blood and liver titanium oxide NPs. It took several days after the oral injection of mice to find the cowpea mosaic virus, a relatively non-pathogenic nanometer-sized plant virus, in numerous tissues throughout the body. These included the spleen and kidneys, liver and lymph nodes, the lungs and small intestines, the brain, and bone marrow. Particles ingested *via* intestinal absorption of dental prosthetics porcelain were first removed by the liver before entering blood and kidneys in a case study. Because of this, it is not clear which organs and tissues will be translocated first.

ADVERSE HEALTH EFFECTS OF GASTROINTESTINAL TRACT UPTAKE

Reaction Reduced Toxicity

When ingested particles interact with chemicals in the digestive tract, they may reduce toxicity. When used *in vitro* for conditions including Crohn's disease, ulcerative colitis, and even cancer, it is less cytotoxic because of the higher protein concentration. In persons with cancer, Crohn's disease, and ulcerative colitis, colloidal gold NPs have been identified. However, healthy people did not contain these particles in their tissues. The afflicted participants' NPs had a wide range of chemical compositions. However, none of them was dangerous when taken as a whole. It was found that carbon, ceramic phyllosilicates, and gypsum were present in the colon mucosa as sulfur, calcium, silicon, stainless steel, and silver. Near the border between healthy and cancerous tissue, the particles were found. There were particles as small as 50 nm and as large as 100 nm, and the smaller ones penetrated deeper. Particles smaller than 20 nm have been shown in these tests to be unsuccessful at crossing the gastrointestinal barrier.

Both natives and immigrants from undeveloped nations are susceptible to Crohn's disease, which affects most developed world's populations. It affects one in every thousand people. Genetic predisposition and environmental variables are to blame for the development of Crohn's disease. High levels of dietary NPs (100–1 nm) have lately been suggested to be connected to Crohn's disease. It was shown that macrophages in the human stomach had exogenous NPs, the first sign of Crohn's disease lesions.

Tiny anatase TiO_2 particles, such as those found in food additives, flaky silicate from natural clay, and other types of ambient silicates, enter the lymphoid tissue of macrophages. The findings were confirmed by microscopy studies. Crohn's disease symptoms appear to be alleviated by exogenous particles. An aberrant reaction to dietary NPs, rather than overconsumption, may be a fault, according to some specialists. More specifically, some people may have a genetic predisposition to being influenced by NPs and developing Crohn's disease. According to some evidence, dietary NPs may worsen Crohn's disease inflammation. The intake of dietary particles was measured in these investigations.

Nonetheless, they did not examine outdoor and indoor nanoparticle pollution levels in the subjects' homes. The mucociliary escalator clears large amounts of NPs, which are then swallowed and eventually reach the gastrointestinal system.

Crohn's disease and ulcerative colitis, both thought to be caused by gastrointestinal nanoparticle absorption, have no known cures and must typically be treated surgically. In order to maintain a state of remission, patients take anti-inflammatory medications and consume specially-made liquid meals. NPs in the diet have been connected to the development of chronic illnesses. As a result, they should be avoided or subject to stricter regulation in the food supply chain [12].

Dermal Uptake of Nps

There are a variety of ways to provide medication, including inhalation and absorption through the skin or digestive system. Chemical absorption *via* the skin must be taken into account when evaluating risk. A good example is the skin, which makes up more than ten percent of the body's total mass and performs several functions, such as protection, homeostasis maintenance, metabolic synthesis, and deposition. Based on the physicochemical properties of the molecule, hair follicles and sweat glands have been identified as potential penetration sites. Passive skin absorption of small lipophilic compounds has long been documented. Dermatological conditions such as contact dermatitis, atopic eczema, and necrotizing dermatitis, as well as the integrity of the skin barrier, may all affect dermal absorption. Skin absorption may also be improved by mechanical flexions, irritating detergents, and chemicals. Particle skin penetration has been studied sparingly, with inconsistent results due to varying methodologies and procedures, laboratory settings, or a lack of uniform evaluation protocols. The respiratory route into the body must be considered as a further problem. Skin is also considered less porous; hence the danger perception is lower when utilizing this method. According to research published in the literature, NPs can enter our bodies through many routes, including the skin.

Polystyrene NPs were exposed to porcine skin for 0.5, 1, and 2 hours in vertical diffusion cells. Observations of the follicular apertures indicated an accumulation of time-dependent polystyrene NPs. The smaller particle sizes facilitated the follicular localization. As Tinkle and colleagues discovered, mechanical flexion helped fluorescent dextran micrometer-sized particles penetrate deeper into the dermal layers while testing its effects on skin nanoparticle uptake in normal conditions. Concerns concerning immunomodulation arose when it was discovered that NPs injected into the dermis traveled to nearby lymph nodes through skin macrophages and Langerhans cells [40].

Penetration Sites

The epidermis, dermis, and subcutaneous layers are the three layers of the skin that make up the whole surface of the body. The stratum corneum, the skin's outermost layer, makes it more difficult for ionic chemicals and water-soluble molecules to get through. There are sweat glands, sebaceous glands, and hair follicles in the pores of the epidermis. As with many other nanoparticle-related issues, there is an ongoing debate over dermal penetration. NPs can permeate the stratum corneum, according to several studies. Often, nanoparticle penetration occurs in hair follicles and skin stretched or damaged. NPs have been proven to enter intracellularly in cell culture tests. Human epidermal keratinocytes absorb

MWCNTs and release pro-inflammatory mediators into the cytoplasmic vacuoles of keratinocytes.

More than 2400 meters and 2.5 millimeters may be penetrated by spherical particles that have sizes between 750 nanometers and 6 micrometers. Particles having a diameter bigger than 500 nanometers may penetrate the skin barrier and enter the body.

When the skin is flexed, NPs have been found to permeate the skin, whereas stationary skin is impenetrable. As a result, mechanical deformation of the stratum corneum can transport particles into the epidermis and dermis *via* the stratum corneum.

It is debatable whether or not the TiO_2 NPs used in commercial sunscreens make it to the skin. NPs are allowed to penetrate at low concentrations. An 8 percent sunscreen using NPs in the 10–15 nm range failed to permeate human skin. Oil-in-water emulsions, on the other hand, penetrated more deeply into hairy skin and pores. Sunscreen in each hair follicle is less than 1% of the applied amount.

ADVERSE HEALTH EFFECTS OF DERMAL UPTAKE

Since workers are regularly exposed to microscopic particles and short fiber lengths during numerous manufacturing processes, it is safe to assume that many of these operations pose an occupational safety and health hazard. Beryllium sensitivity in employees exposed to nanoparticulate beryllium while wearing inhalation protection equipment may have arisen. In addition to latex sensitivity, this could apply to other materials that induce skin responses, such as latex. Fig. (**4d**) and (**4f**) illustrate that lymphatic system uptake of NPs through the dermis causes podoconiosis and Kaposi's sarcoma. The toxicology of titanium dioxide, a common ingredient in sunscreen, has raised concerns. It is tin oxide. Cosmetic-derived titanium dioxide toxicity is a hotly debated subject right now. Skin-damaging UVB and UVA light rays are reflected and scattered by the cloud, including those with wavelengths between 290 and 320 nanometers (nm). Additionally, a considerable amount of the UV light that strikes TiO_2 is absorbed, which leads to aqueous media of a wide range of reactive oxygen species, such as superoxide anion radicals and hydrogen peroxide.

These reactive oxygen radicals have the potential to cause significant DNA damage. Exposing cultured human bladder cancer cells to UV light irradiation, which creates reactive oxygen species generated by titanium dioxide particles, may slow the development of those cells. Studies have shown that the titanium dioxide particles in sunscreen cause *In vitro* and *in vivo* DNA damage.

Fig. (4). (**a**) The eruption plume of St. Helen volcano, in 1980. (**b**) Rabaul Eruption Plume, New Britain Island, 1994. The large scale of eruption can be compared to the Earth's curvature; (**c**) Scanning electron microscope image of volcanic ash from the first volcanic eruption of Mount St. Helens, Washington State, the USA, in 1980. (**d**) Podoconiosis. (**e**) Volcanic iron oxide-rich soil in Rwanda. (**f**) Aggressive African-endemic Kaposi's sarcoma of the foot [12].

There are conflicting reports on the toxicity of titanium dioxide NPs when exposed to light without UV light. When administered orally, NPs had no inflammatory or genotoxic effects on rats. Titanium dioxide installation has been linked to long-term rat lung inflammation and pro-inflammatory effects on human endothelial cell cultures in the lab. Silver. In the early stages of wound healing, silver is known to have an antimicrobial impact, lowering inflammation and aiding recovery. Silver NPs and ion cytotoxicity have been studied in numerous laboratories worldwide, with contradicting results. Despite its antibacterial properties, silver may be harmful to human cells. For both keratinocytes and fibroblasts, silver at concentrations fatal to bacteria also kills them [12].

NPs Uptake *via* Injection

The use of NPs in drug delivery has been researched. Subcutaneous tissue, muscle, blood arteries, or internal organs may receive fluid *via* injection. Following injection, the movement of NPs is influenced by the injection site: After quickly moving through the circulatory system *via* intravenous administration, NPs injected intradermally are taken up by lymph nodes, whereas intramuscular injections lead to lymph nodes being taken up by neuronal and lymphatic systems. This synaptic uptake, for example, was observed in mice injected with tiny magnetic NPs (less than 100 nm). When given intravenously, NPs have a much longer half-life than when taken orally. 90% of the functionalized fullerenes injected one week ago are still present. Fullerenes, quantum dots, polystyrene, and plant viruses have all been detected in the liver, bone marrow, and lymph nodes, as well as the small intestines and brain. Drug addicts' livers have been shown to contain talc particles.

The surface features and size of particles affect their dispersion in the body. Surfactant-coated NPs significantly impact their dispersion in the body prior to injection.

Hepatic and splenic localization is virtually impossible when polyethylene glycol or other compounds are used as a coating. Adding cationic compounds to the nanoparticle's surface, which increases arterial absorption tenfold, is another example. Injecting NPs into the body may pose health risks because of their chemical and electrical properties. Administering NPs intravenously often results in hypersensitivity as a side effect. The activation of the complement system is assumed to be the source of this reaction in many individuals [12].

CONCLUSION

The safety issues derived from NPs routes of entry and their potential bio-distribution are governed by surface area, shape, agglomeration, aggregation solubility and size with protein (opsonization) interactions within the host. The size fractions in the nanoscale range have greater lung deposition and rapid systemic translocation having various inflammatory, oxidative and cytotoxic effects on experimental animals than larger particles. With these discussed possible potential routes of NPs, nanotechnology research should proceed with caution. The combination of hazard and production should go hand in hand to reduce the potential acquisition of NPs through good manufacturing practice (GMP), good laboratory practice (GLP) and International Standards Organisation

(ISO). Suitable quality control procedures should be part of the process to ensure NPs product safety and quality and hence part of the company quality assurance scheme. Also, the manufacturing industries of nanotechnology should work hand in hand with the health and hazard risk assessment to establish a lower health risk of any type emanating from the production and use of NPs. Though there is limited toxicological data available at present, with the current review, there is hope to increase the awareness and safety issues of nanotechnology.

REFERENCES

[1]   P. Laux, J. Tentschert, C. Riebeling, A. Braeuning, O. Creutzenberg, A. Epp, V. Fessard, K.H. Haas, A. Haase, K. Hund-Rinke, N. Jakubowski, P. Kearns, A. Lampen, H. Rauscher, R. Schoonjans, A. Störmer, A. Thielmann, U. Mühle, and A. Luch, "Nanomaterials: certain aspects of application, risk assessment and risk communication", *Arch. Toxicol.,* vol. 92, no. 1, pp. 121-141, 2018.
[http://dx.doi.org/10.1007/s00204-017-2144-1] [PMID: 29273819]

[2]   H. Maeda, "Toward a full understanding of the EPR effect in primary and metastatic tumors as well as issues related to its heterogeneity", *Adv. Drug Deliv. Rev.,* vol. 91, pp. 3-6, 2015.
[http://dx.doi.org/10.1016/j.addr.2015.01.002] [PMID: 25579058]

[3]   M.J. Ernsting, M. Murakami, A. Roy, and S.D. Li, "Factors controlling the pharmacokinetics, biodistribution and intratumoral penetration of nanoparticles", *J. Control. Release,* vol. 172, no. 3, pp. 782-794, 2013.
[http://dx.doi.org/10.1016/j.jconrel.2013.09.013] [PMID: 24075927]

[4]   Q. Mu, G. Jiang, L. Chen, H. Zhou, D. Fourches, A. Tropsha, and B. Yan, "Chemical basis of interactions between engineered nanoparticles and biological systems", *Chem. Rev.,* vol. 114, no. 15, pp. 7740-7781, 2014.
[http://dx.doi.org/10.1021/cr400295a] [PMID: 24927254]

[5]   D. Peer, "Immunotoxicity derived from manipulating leukocytes with lipid-based nanoparticles", *Adv. Drug Deliv. Rev.,* vol. 64, no. 15, pp. 1738-1748, 2012.
[http://dx.doi.org/10.1016/j.addr.2012.06.013] [PMID: 22820531]

[6]   C. Auría-Soro, T. Nesma, P. Juanes-Velasco, A. Landeira-Viñuela, H. Fidalgo-Gomez, V. Acebes-Fernandez, R. Gongora, M.J. Almendral Parra, R. Manzano-Roman, and M. Fuentes, "Interactions of nanoparticles and biosystems: Microenvironment of nanoparticles and biomolecules in nanomedicine", *Nanomaterials (Basel),* vol. 9, no. 10, p. 1365, 2019.
[http://dx.doi.org/10.3390/nano9101365] [PMID: 31554176]

[7]   J.M. Zook, R.I. MacCuspie, L.E. Locascio, M.D. Halter, and J.T. Elliott, "Stable nanoparticle aggregates/agglomerates of different sizes and the effect of their size on hemolytic cytotoxicity", *Nanotoxicology,* vol. 5, no. 4, pp. 517-530, 2011.
[http://dx.doi.org/10.3109/17435390.2010.536615] [PMID: 21142841]

[8]   C. Yan, and T. Wang, "A new view for nanoparticle assemblies: from crystalline to binary cooperative complementarity", *Chem. Soc. Rev.,* vol. 46, no. 5, pp. 1483-1509, 2017.
[http://dx.doi.org/10.1039/C6CS00696E] [PMID: 28059420]

[9]   P. Gehr, "Interaction of nanoparticles with biological systems", *Colloids Surf. B Biointerfaces,* vol. 172, pp. 395-399, 2018.
[http://dx.doi.org/10.1016/j.colsurfb.2018.08.023] [PMID: 30195156]

[10]   S.R. Saptarshi, A. Duschl, and A.L. Lopata, "Interaction of nanoparticles with proteins: relation to bio-reactivity of the nanoparticle", *J. Nanobiotechnology,* vol. 11, no. 1, p. 26, 2013.
[http://dx.doi.org/10.1186/1477-3155-11-26] [PMID: 23870291]

[11]   S. Saallah, and I.W. Lenggoro, "Nanoparticles carrying biological molecules: Recent advances and

applications", *Kona Powder Particle J.,* vol. 35, no. 0, pp. 89-111, 2018.
[http://dx.doi.org/10.14356/kona.2018015]

[12]   C. Buzea, I.I. Pacheco, and K. Robbie, "Nanomaterials and nanoparticles: Sources and toxicity",
       Biointerphases, vol. 2, no. 4, pp. MR17-MR71, 2007.
       [http://dx.doi.org/10.1116/1.2815690] [PMID: 20419892]

[13]   B. Singh, and S. Mitragotri, "Harnessing cells to deliver nanoparticle drugs to treat cancer", *Biotechnol
       Adv.,* vol. 42, p. 107339, 2020.
       [http://dx.doi.org/10.1016/j.biotechadv.2019.01.006] [PMID: 30639928]

[14]   A. Hayes, and S. Bakand, "Inhalation toxicology", *EXS.,* vol. 100, pp. 461-488, 2010.
       [http://dx.doi.org/10.1007/978-3-7643-8338-1_13] [PMID: 20358692]

[15]   B. Siddhardha, M. Dyavaiah, and K. Kasinathan, *Model Organisms to Study Biological Activities and
       Toxicity of Nanoparticles.* Springer, 2020.
       [http://dx.doi.org/10.1007/978-981-15-1702-0]

[16]   S.D. Conner, and S.L. Schmid, "Regulated portals of entry into the cell", *Nature,* vol. 422, no. 6927,
       pp. 37-44, 2003.
       [http://dx.doi.org/10.1038/nature01451] [PMID: 12621426]

[17]   P. Watson, A.T. Jones, and D.J. Stephens, "Intracellular trafficking pathways and drug delivery:
       fluorescence imaging of living and fixed cells", *Adv Drug Deliv Rev.,* vol. 57, no. 1, pp. 43-61, 2005.
       [http://dx.doi.org/10.1016/j.addr.2004.05.003]

[18]   J. Chang, Y. Jallouli, M. Kroubi, X. Yuan, W. Feng, C. Kang, P. Pu, and D. Betbeder,
       "Characterization of endocytosis of transferrin-coated PLGA nanoparticles by the blood–brain
       barrier", *Int. J. Pharm.,* vol. 379, no. 2, pp. 285-292, 2009.
       [http://dx.doi.org/10.1016/j.ijpharm.2009.04.035] [PMID: 19416749]

[19]   M. Walsh, M. Tangney, M.J. O'Neill, J.O. Larkin, D.M. Soden, S.L. McKenna, R. Darcy, G.C.
       O'Sullivan, and C.M. O'Driscoll, "Evaluation of cellular uptake and gene transfer efficiency of
       pegylated poly-L-lysine compacted DNA: implications for cancer gene therapy", *Mol. Pharm.,* vol. 3,
       no. 6, pp. 644-653, 2006.
       [http://dx.doi.org/10.1021/mp0600034] [PMID: 17140252]

[20]   G. Bao, and X.R. Bao, "Shedding light on the dynamics of endocytosis and viral budding", *Proc. Natl.
       Acad. Sci. USA,* vol. 102, no. 29, pp. 9997-9998, 2005.
       [http://dx.doi.org/10.1073/pnas.0504555102] [PMID: 16009932]

[21]   M. Ehrlich, W. Boll, A. van Oijen, R. Hariharan, K. Chandran, M.L. Nibert, and T. Kirchhausen,
       "Endocytosis by random initiation and stabilization of clathrin-coated pits", *Cell,* vol. 118, no. 5, pp.
       591-605, 2004.
       [http://dx.doi.org/10.1016/j.cell.2004.08.017] [PMID: 15339664]

[22]   L. Pelkmans, and A. Helenius, "Endocytosis *via* caveolae", *Traffic,* vol. 3, no. 5, pp. 311-320, 2002.
       [http://dx.doi.org/10.1034/j.1600-0854.2002.30501.x] [PMID: 11967125]

[23]   Z. Wang, C. Tiruppathi, R.D. Minshall, and A.B. Malik, "Size and dynamics of caveolae studied using
       nanoparticles in living endothelial cells", *ACS Nano,* vol. 3, no. 12, pp. 4110-4116, 2009.
       [http://dx.doi.org/10.1021/nn9012274] [PMID: 19919048]

[24]   S. Tzlil, M. Deserno, W.M. Gelbart, and A. Ben-Shaul, "A statistical-thermodynamic model of viral
       budding", *Biophys. J.,* vol. 86, no. 4, pp. 2037-2048, 2004.
       [http://dx.doi.org/10.1016/S0006-3495(04)74265-4] [PMID: 15041646]

[25]   D. Effenterre, and D. Roux, "Adhesion of colloids on a cell surface in competition for mobile
       receptors", *Europhys. Lett.,* vol. 64, no. 4, pp. 543-549, 2003.
       [http://dx.doi.org/10.1209/epl/i2003-00268-x]

[26]   S. Zhang, J. Li, G. Lykotrafitis, G. Bao, and S. Suresh, "Size-dependent endocytosis of nanoparticles",
       Adv. Mater., vol. 21, no. 4, pp. 419-424, 2009.

[http://dx.doi.org/10.1002/adma.200801393] [PMID: 19606281]

[27]   Y. Aoyama, "Artifical viruses and their application to gene delivery. Sizecontrolled Nanoparticles., gene coating with glycocluster", *J. Am. Chem. Soc.,* vol. 125, pp. 3455-3457, 2003.
[http://dx.doi.org/10.1021/ja029608t] [PMID: 12643707]

[28]   P. Decuzzi, and M. Ferrari, "The role of specific and non-specific interactions in receptor-mediated endocytosis of nanoparticles", *Biomaterials,* vol. 28, no. 18, pp. 2915-2922, 2007.
[http://dx.doi.org/10.1016/j.biomaterials.2007.02.013] [PMID: 17363051]

[29]   H. Gao, W. Shi, and L.B. Freund, "Mechanics of receptor-mediated endocytosis", *Proc. Natl. Acad. Sci. USA,* vol. 102, no. 27, pp. 9469-9474, 2005.
[http://dx.doi.org/10.1073/pnas.0503879102] [PMID: 15972807]

[30]   O. Harush-Frenkel, N. Debotton, S. Benita, and Y. Altschuler, "Targeting of nanoparticles to the clathrin-mediated endocytic pathway", *Biochem. Biophys. Res. Commun.,* vol. 353, no. 1, pp. 26-32, 2007.
[http://dx.doi.org/10.1016/j.bbrc.2006.11.135] [PMID: 17184736]

[31]   P. Nativo, I.A. Prior, and M. Brust, "Uptake and intracellular fate of surface-modified gold nanoparticles", *ACS Nano,* vol. 2, no. 8, pp. 1639-1644, 2008.
[http://dx.doi.org/10.1021/nn800330a] [PMID: 19206367]

[32]   M. Gou, X. Zheng, K. Men, J. Zhang, L. Zheng, X. Wang, F. Luo, Y. Zhao, X. Zhao, Y. Wei, and Z. Qian, "Poly(epsilon-caprolactone)/poly(ethylene glycol)/poly(epsilon-caprolactone) nanoparticles: preparation, characterization, and application in doxorubicin delivery", *J. Phys. Chem. B,* vol. 113, no. 39, pp. 12928-12933, 2009.
[http://dx.doi.org/10.1021/jp905781g] [PMID: 19736995]

[33]   S. Hong, A.U. Bielinska, A. Mecke, B. Keszler, J.L. Beals, X. Shi, L. Balogh, B.G. Orr, J.R. Baker Jr, and M.M. Banaszak Holl, "Interaction of poly(amidoamine) dendrimers with supported lipid bilayers and cells: hole formation and the relation to transport", *Bioconjug. Chem.,* vol. 15, no. 4, pp. 774-782, 2004.
[http://dx.doi.org/10.1021/bc049962b] [PMID: 15264864]

[34]   Y. Roiter, M. Ornatska, A.R. Rammohan, J. Balakrishnan, D.R. Heine, and S. Minko, "Interaction of nanoparticles with lipid membrane", *Nano Lett.,* vol. 8, no. 3, pp. 941-944, 2008.
[http://dx.doi.org/10.1021/nl080080l] [PMID: 18254602]

[35]   H.Y. Nam, S.M. Kwon, H. Chung, S.Y. Lee, S.H. Kwon, H. Jeon, Y. Kim, J.H. Park, J. Kim, S. Her, Y.K. Oh, I.C. Kwon, K. Kim, and S.Y. Jeong, "Cellular uptake mechanism and intracellular fate of hydrophobically modified glycol chitosan nanoparticles", *J. Control. Release,* vol. 135, no. 3, pp. 259-267, 2009.
[http://dx.doi.org/10.1016/j.jconrel.2009.01.018] [PMID: 19331853]

[36]   A. Kumari, and S.K. Yadav, "Cellular interactions of therapeutically delivered nanoparticles", *Expert Opin Drug Deliv.,* vol. 8, no. 2, pp. 141-151, 2011.
[http://dx.doi.org/10.1517/17425247.2011.547934] [PMID: 21219249]

[37]   E. Panzarini, S. Mariano, E. Carata, F. Mura, M. Rossi, and L. Dini, "Intracellular transport of silver and gold nanoparticles and biological responses: An update", *Int. J. Mol. Sci.,* vol. 19, no. 5, p. 1305, 2018.
[http://dx.doi.org/10.3390/ijms19051305] [PMID: 29702561]

[38]   K. Sawicki, M. Czajka, M. Matysiak-Kucharek, B. Fal, B. Drop, S. Męczyńska-Wielgosz, K. Sikorska, M. Kruszewski, and L. Kapka-Skrzypczak, "Toxicity of metallic nanoparticles in the central nervous system", *Nanotechnol. Rev.,* vol. 8, no. 1, pp. 175-200, 2019.
[http://dx.doi.org/10.1515/ntrev-2019-0017]

[39]   Y.N. Zhang, W. Poon, A.J. Tavares, I.D. McGilvray, and W.C.W. Chan, "Nanoparticle–liver interactions: Cellular uptake and hepatobiliary elimination", *J. Control. Release,* vol. 240, pp. 332-348, 2016.

[http://dx.doi.org/10.1016/j.jconrel.2016.01.020] [PMID: 26774224]

[40]   M. Crosera, M. Bovenzi, G. Maina, G. Adami, C. Zanette, C. Florio, and F. Filon Larese, "Nanoparticle dermal absorption and toxicity: a review of the literature", *Int. Arch. Occup. Environ. Health,* vol. 82, no. 9, pp. 1043-1055, 2009.
[http://dx.doi.org/10.1007/s00420-009-0458-x] [PMID: 19705142]

CHAPTER 4

Protein–Nanoparticles Interactions

Abstract: Large surface area, small size, strong optical properties, controllable structural features, variety of bioconjugation chemistries, and biocompatibility make many different types of nanoparticles (NPs), such as gold NPs, useful for many biological applications, such as biosensing, cellular imaging, disease diagnostics, drug delivery, and therapeutics. Recently, interactions between proteins and NPs have been extensively studied to understand, control, and utilize the interactions involved in biomedical applications of NPs and several biological processes, such as protein aggregation, for many diseases, including Alzheimer's. These studies also offer fundamental knowledge on changes in protein structure, protein aggregation mechanisms, and ways to unravel the roles and fates of NPs within the human body.

Keywords: Nanoparticles interactions, Protein interactions, Protein structure.

INTRODUCTION

In biology, it is a (near) universal law that material is always covered with proteins when it comes into contct with a physiological environment. Much of the bionanoscience world will be better understood if these phenomena can be explained. The particle's "corona" of serum and other physiological fluid proteins, rather than the particle itself, is the effective unit of attention in cell nanomaterial interactions, according to a recent hypothesis. Understanding that live cells read more than just the protein layer's content and organization is crucial. They also take note of how quickly the proteins on the NPs swap. There is a wide range of protein binding affinity on a particle; hence the nanoparticle surface reveals a variety of protein residence times. There is not a solid layer present; instead, a protein-linked 'corona' is seen. As a result of the wide variety of binding strategies (one for each protein), we may anticipate a wide range of equilibrium constants (some of which are quite competitive).

Surface protein concentration and interaction rate will be high on nanoparticle surfaces. In order to gain an understanding of how a specific nanoparticle's protein corona varies over time, we may evaluate the kinetic on/off rates of each protein in the plasma. This corona may not be able to reach equilibrium promptly when exposed to biological fluid. There is a need to replace proteins with low concen-

Seyed Morteza Naghib and Hamid Reza Garshasbi

tration, slower exchange, and higher affinity with those that dissolve fast in the presence of other proteins. There may be a need for exchange systems when particles move between different compartments or organs, such as the circulation into cells or the movement of the cytosol to the nucleus. Because of this, the cell perceives and interacts with the protein corona as its biological self. Adsorption of bovine serum albumin (BSA), myoglobin (Mb), and cytochrome C (Cyt C) onto Au NPs is characterized by an irreversibly adsorbed portion and a proportion of protein adsorption that may be reversed. NPs and proteins are reported to have a hard corona, which has a long residence time and has strong adsorption to NPs.

Shorter residence times or lower affinities are common among the proteins that make up an easier-to-fold corona. According to a literature review on nanoparticle-protein binding, most nanoparticle types investigated too far bind apo lipoproteins. At first look, this seems to be a completely unexpected outcome. Although apolipoproteins are found in lipoprotein complexes at the nanoscale, this proves that they are a component of lipoprotein complexes in general.

Several NPs and the apolipoprotein E molecule have a role in nanoparticle transport in animals and humans. Due to their role in cholesterol metabolism, lipoprotein complexes are crucial to NPs interaction with cells. This means that NPs may enter cells through receptors on the cell surface that recognize apolipoprotein complexes that have been surface-adsorbed. As apolipoprotein E is important in the trafficking of cells to the brain, this might have substantial repercussions for neurotoxicity and neurotherapeutics development. The nanoparticle-protein corona may impact both the ultimate subcellular position of a nanoparticle after it interacts with a cell and the range of disease processes it may access (in addition to size and shape).

The biomolecule corona, which facilitates nanoparticles' interface with cellular machinery, should be used in the future to classify NPs. Toxicology at the nanoscale might be revolutionized, as could nanomedicine delivery systems. It has been shown for the first time that protein adsorption to NPs has a direct biological impact. Single-walled carbon nanotubes (SWNTs) and albumin-coated amorphous silica reduced lipopolysaccharide-induced activation of Cyclooxygenase-2 (Cox-2) in macrophages. Nonionic surfactants are used to prevent albumin adsorption, which has anti-inflammatory characteristics, on the NPs.

The presence of adsorbed proteins is suggested to modulate the detrimental effects of SWNTs and nanoscale amorphous silica. Despite this, it is unclear whether albumin would attach to NPs in other binding contexts, such as a plasma or the inside of a cell [1].

FORMATION OF THE PROTEIN CORONA

More and more people are seeing how nanotechnology's ability to make ever-smaller NPs while simultaneously improving control over their physical and chemical properties has propelled the ability to do so. These NPs are more reactive due to an increase in the number of atoms on their surfaces. Targeted medication delivery and imaging may be accomplished *via* intravenous injection of NPs developed for biological uses like this. To create the "protein corona," NPs quickly get coated with certain blood plasma proteins to reach the biological environment (plasma, for example). To reach plasma proteins through other routes of exposure, NPs must first cross the body's physiological barriers (the skin, gastrointestinal system, and the lungs, for example), collecting up biomolecules along the way.

As a result of the protein-coated particles' ability to expose new epitopes, alter the adsorbed protein conformation (either temporarily or permanently), alter protein function, and possibly even cause avidity effects due to the close spatial repetition of the same protein, a wide range of biological effects are possible. One layer is made up of proteins with long half-lives and rapid exchange rates with free proteins, while the other layer is made up of proteins with shorter half-lives and a slower exchange rate with free proteins, as revealed by researchers (the soft corona). Hard coronas have a longer life span. The hard corona, rather than the smooth NP surface, is now thought to interact with cell receptors and impact NP fate. To ensure the safety of nanomedicines and consumer products, it is essential to establish the proteins that adsorb on the NP surface and their respective lifetimes and conformations.

Biomedical applications may benefit from developing "safe by design" NPs if the interactions between NPs and proteins can be better understood and predicted. Many different qualities of the NP and the adsorbing proteins may affect the corona's appearance, including its physical and chemical properties.

The hard corona of protein-coated NPs has a substantial biological impact because of their high affinity (*i.e.,* the capacity to arrange themselves in a certain manner to trigger a specific receptor response). Proteins with high abundance may initially bind but are rapidly replaced by proteins with lower quantities and greater affinity. Since the timescales of cellular processes are so short, it is crucial to evaluate protein lifetimes on the surface of NPs to determine NP biological response. NP interactions with living matter are believed to be mediated by proteins with a more tightly attached corona (hard corona). The structure and affinity of proteins adsorbed on NP surfaces must be well elucidated. To do this characterization, proteins must be separated from the NPs' surfaces. It is difficult to isolate the nanoparticle-protein complex process. Plasma proteins compete for binding in kinetic and thermodynamically controlled systems.

Soft and hard coronas are being studied to understand better how this occurs. The most common method for removing low-binding proteins is centrifugation and washing of NPs. Physicochemical approaches for eliminating protein-NP complexes from unattached proteins are acceptable and helpful using DCS, which produces almost equivalent results in plasma and after centrifugation and washing of the plasma. Differentiating between proteins with varying degrees of affinity has been made possible by developing innovative techniques that do not break protein-NP complexes or provide new binding possibilities. Size-exclusion chromatography, isothermal titration calorimetry (ITC), and surface-plasmon resonance (SPR) are ways to study the interactions between NPs and different types of proteins.

With four different NIPAM/N-tert-butyl acrylamide copolymer NPs with increasing hydrophobicity on the surface, ITC can be used to figure out how many proteins are attached to the surface of the NPs. Due to the need for an enthalpic shift in the NP-protein interaction and that quantitative information can only be collected from single protein-NP studies, this method's thermodynamic and stoichiometric information is restricted.

Size-exclusion chromatography and thiol-linked gold chip SPR were employed to analyze the NP surface's protein exchange rates and affinity. SEC may be less disruptive to protein and particle complexes, as shown by this early study of NP-plasma interactions. This study's last phase included measuring the characteristics of the NP surface protein and the plasma protein exchange rates using SEC-gel filtration. It has been established that a precise centrifugation and washing technique may yield equivalent high-reproducibility coronas in subsequent studies. In the SEC-gel filtering process, higher yields may be attained by adjusting the protein/NP ratio and the size of the column. The bound proteins can then be detected using mass spectrometry methods.

Protein coronas surround magnetic NPs, first described by Mahmoudi *et al.* Quantitative and qualitative evaluations are facilitated using soft and hard coronas (*e.g.,* protein conformation and geometry). SPIONs (super-paramagnetic iron oxide NPs) have been chosen above other magnetic NPs because of their distinct magnetic properties and exceptional biocompatibility.

As part of this method, once the NPs had interacted with proteins for the required time, they were stuck in a magnetic column utilizing magnetic-activated cell sorting. Proteins that had not been bound to NPs were found in the flow-through fraction. Protein-washing solutions of various molarities were used to wash the trapped NPs to test their affinity for protein-SPION binding (*e.g.,* KCl).

THERMODYNAMIC AND KINETIC ASPECTS OF THE NP_PROTEIN CORONA

Only by understanding the thermodynamics and kinetics of corona formation can the composition of the NP-protein corona and its implications for the environment or biology be fully grasped and appreciated. In order for NPs to engage with biological surfaces and cell receptors, they must compete with other NPs and exchange free proteins in the media. For protein-NP interactions, it is possible to use ITC to estimate stoichiometric parameters such as affinity and enthalpy. Even though it is tough to put all the pieces together to see how the soft corona is generated and traded, the medical device industry has many data on protein binding to surfaces. In order to accurately anticipate nanoparticle destiny in biological fluids, one must understand the dynamics of protein adsorption on nanoparticle surfaces. Protein surface affinity may be affected by the Vroman effect, which is dependent on protein concentrations and diffusion coefficients on the available surface area. An affinity-reduced protein replaces the original high-affinity protein at the site of protein adsorption on a surface. On the other hand, NPs exhibit different behavior due to their high surface curvature and enormous surface area.

Depending on how the various parts combine, different NPs may have different effects. Polystyrene NPs with various functional groups (COOH, CH_3) may temporarily adsorb low-affinity proteins (such as albumin), which are replaced by fibrinogen. Because of fibrinogen's affinity, this happens. Sulfonated polystyrene nanoparticle affinity for fibrinogen diminishes dramatically with increasing plasma concentration, indicating a role for protein concentration to NP surface area ratio. There is no change in the surface area of particles [2].

NPS INDUCE CHANGES IN THE STRUCTURE OF ADSORBED PROTEINS

In this part, we will discuss what happens to the proteins that are bound to the surface of the NP. The adsorbed protein's structure and function may be altered by the NP surface, which might have an impact on the NP's overall bioactivity. Adsorbed protein molecules benefit from the increased flexibility and surface area of curved NP surfaces. Coiled NP surfaces have been shown to alter protein secondary structures permanently in a few rare circumstances. Chemical characteristics and structural flexibility of specific proteins influence secondary structure alterations on the surface. Gold NPs caused a dose-dependent increase in structural changes in bovine serum albumin (BSA).

As a result, BSA's structure did not alter when attached to carbon C60 fullerene NP. A crucial cytoskeletal protein, tubulin polymerization, was reduced by Titanium dioxide (TiO_2) NP-induced conformational changes. ZnO-NP interaction with BSA was studied, but no structural alterations were discovered. Minor modifications in conformation, on the other hand, were discovered. SPIONs irreversibly altered the secondary structure of transferrin when they interacted with the protein. NP-induced protein conformational changes may alter enzyme DNA transcription and protein-protein interactions. Due to the NP surface's effect on the active site's conformation, enzyme activity may decline.

Two different studies found that SWCNTs reduced enzyme structure and catalytic activity. Researchers found that silica NP, RNAse, and lysozyme retained their original forms after exposure to UV rays. On the other hand, Albumin and lactoperoxidase suffered a permanent conformational change. Similarly, the enzyme's active site might be more accessible to its substrate if it undergoes a conformational shift. Human carbonic anhydrase's structural shift resembled that of a molten globule after exposure to silica NPs. Three intermediate native-like conformations with catalytic activity were formed when the nucleotide NP was removed.

Horseradish peroxide, subtilisin Carlsberg, and chicken egg white lysozyme were unaffected by denaturation. The unique folding of the protein polypeptide chain creates conformational epitopes. Anomalies in protein unfolding caused by the NP surface might lead to the discovery of new conformational epitopes, or native protein structure unfolding could disclose previously undiscovered epitopes (Fig. **1**). For example, occult epitopes may cause an undesirable immune response to the attached proteins. However, epitopes with a length of 10–12 amino acids on the protein's core structure may also elicit an immunological response. By activating the Mac-1 receptor on THP-1 cells, Deng and colleagues found that the NF-kβ pathway is activated when positive-charged poly (acrylic acid)-conjugated gold NPs bind and unfold fibrinogen in blood plasma and activate the receptor Mac-1. A lack of tolerance to one's proteins due to structural changes might trigger an autoimmune reaction, a major cause for worry. Fibrils may arise as a result of NP-induced changes in protein structure. It has been shown that the NPs copolymer, cerium, carbon nanotubes, and quantum dots, among others, may cause cancer and promote fibrillation of 2-microglobulin, resulting in a rise in protein localization on the NP surface and the creation of oligomer. Parkinson's and Alzheimer's disease are linked to protein fibrillation.

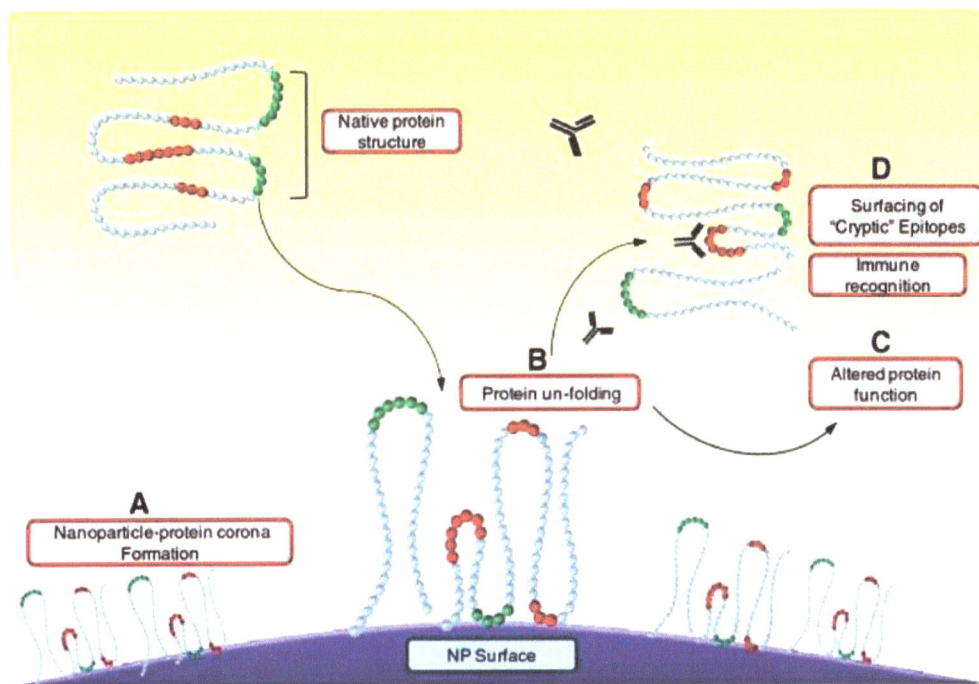

Fig. (1). Diagram showing how NP causes the protein molecule to unfold and what this means. (**A**) Protein molecules stick to the surface of the NP, making a (**B**) NP-PC complex. The surface of the NP may change the natural shape of the protein molecule that has been absorbed, causing it to unfold. There are many reasons why the shape of a protein might change. (**C**) change how the native protein molecule works or even (**D**) expose "cryptic" epitopes, which may cause the immune system to recognize the complex [3].

There may be thermodynamic instability in an adsorbed protein, which makes it more susceptible to chemical denaturation. There is a need for further study on NPs as a platform for protein structural change. The ToxR protein from Vibrio cholera has its periplasmic domain unraveled using ZnO NPs, making it sensitive to chaotropic medicines.

ZnO NPs stabilized lysozyme's -helical content against denaturing agents, which was unexpected. Because of this, the NP surface's chemical properties determine the destiny of proteins that connect to it.

NANOPARTICLE-PROTEIN CORONA: IMPLICATION ON CELLULAR INTERACTIONS

Since they are so small, NPs tend to interact with cells and penetrate membrane barriers in an organism since they are more accessible. If they are less than 100 nm, NPs may penetrate cells. For those smaller than 40 nm, it is possible to penetrate the nucleus, while those less than 35 nm may cross the blood-brain barrier. To absorb NPs, either phagocytsis, macropinocytosis, or endocytosis of the substance occurs. Lysosomes, intracellular vacuoles, and the cytoplasm are all possible destinations for NPs after they are absorbed (as seen with a copolymer NP). The two most important consequences of cellular absorption of NP are cytotoxicity and immunological regulation. In particular, NPs that dissolve in the cell's acidic lysosomal compartments contribute to cellular toxicity. Therefore, this is particularly crucial for them. Investigating every aspect of NP absorption is crucial to fully comprehend the ultimate fate of these NPs inside living systems. The characteristics of the cells interacting with one other and the physical qualities of NP might impact NP absorption. Varying cell types have been shown to absorb the same nanomaterial at different rates. Protein adsorption on the NP surface might occur in a few seconds. NP contact with cellular structures is thus assumed to be indirect. The NP-PC, rather than the naked NP surface, is the major conduit for this phenomenon.

A protein's ability to enter the cell depends on whether the protein's structure is preserved or unfolded before reaching cell surface receptors. Thus, the NP-PC may influence the cell's ability to absorb the NP. In case of physiologically active proteins, this is especially relevant. It has been shown that NP may be absorbed in serum proteins *in vitro*.

Albumin adsorption on single-walled carbon nanotubes (SWCNTs) was found to initiate an anti-inflammatory pathway in RAW macrophages, which suggests that the identity of the protein may affect its bio-reactivity when it is adsorbed on the surface of NPs. The research was carried out by Dutta and colleagues.

Using magnetic NPs pre-coated with BSA as a comparison, researchers found that lung protein SP-A adsorption to magnetic NPs increased macrophage absorption. Albumin was needed on the NPs' surface for caveolae to endocytose fluorescent polystyrene NPs (20-100 nm in size). Polystyrene NPs may be found inside caveolae, the 60-80 nm-diameter cell membrane invaginations, according to an increasing body of research. Apolipoproteins are a kind of protein found in the blood. NP surfaces of all kinds fall within their purview. NP absorption may be boosted by attaching to certain cell receptors, which has drawn the attention of researchers. NPs coated with polysorbate 80 and drug-bound antibodies B and E were shown to be able to pass the blood-brain barrier. Endocytosis mediated by the receptors was proposed in this example. The blood-brain barrier's impenetrable nature prevents the delivery of key medicines and other chemicals to the brain. Even if NPs break past this barrier, there are substantial safety concerns about the toxicity of nanomaterials raised by this development. It is possible that serum proteins, including immunoglobulins and complement system proteins, include poisoning characteristics. If the blood contains too many opsonizing proteins, the NP-protein combination might be mistaken for a foreign object by the immune system, leading to an overreaction. When the macrophages were incubated in serum-enriched media after exposure to NH2-polystyrene NPs in a protein-free medium, they went from clathrin-mediated endocytosis to phagocytosis. Because of this, it may be concluded that serum proteins can greatly impact the absorption of nucleic acid (NP) on its surface.

Additionally, it was shown that the absorption of complement protein C3 and opsonizing protein IgG by murine Kupffer cells was altered by the adsorption of these proteins to lecithin-coated polystyrene NPs of 50 nm.

The hydrodynamic size of the NP is anticipated to increase when proteins are attached to its surface. Nonphagocytic and phagocytic cells may take up NP-protein complexes of this size. Lesniak *et al.* found that polystyrene NP absorption by non-phagocytic lung epithelial cells was greatly increased when the cells were cultured in the non-heated serum. Non-heat-inactivated serum may have a higher absorption because it contains more proteins and heat-labile complement proteins than heat-inactivated serum, as suggested by researchers. It is also plausible that the higher absorption is due to the nonspecific nature of the NP's cell-to-NP interaction, which relies only on protein concentration rather than NP surface specificity. Ehrenberg *et al.* found that NPs were not affected *in vitro*, whether they were treated with whole serum or a serum devoid of several of the most common proteins. According to this study, NPs with protein-bound surfaces are more absorbent, yet the literature has conflicting results. HeLa cells' uptake of FePt NPs was inhibited when the NP-PC was present. Also, lung epithelial cells absorbed silica NPs in serum-free media more effectively than in 10% serum. NP's capacity to absorb seems linked to the quantity and kind of protein on the NP surface. Zeta potential and the NP surface area, two minor Physico-chemical factors, have been demonstrated to influence NP uptake and protein binding in cells.

The absorbance of gold NPs by human breast cells is influenced by their sedimentation capacity, which was not predicted. Oncology cells were exposed to gold NPs in two ways: upright and inverted. It was found by Kim and colleagues that the cell cycle phase had a significant impact on the absorption of 40 nm yellow-green PS-COOH by A549 cells. As cells divide, the NP taken in by parent cells is distributed among the daughter cell, which has ramifications for the clearance or buildup of NP.

In the research on cellular absorption of NPs, many discrepancies exist about the elements that influence this interaction. Bio-reactivity of a nanoparticle (NP) is partly controlled by its surface, which affects how the attached protein's natural structure is perturbed. Most investigations use immortal cell lines, which may have distinct features from their *ex vivo* counterparts. Much of this research has been done in the laboratory. Several factors are at play when extrapolating NPs' *in-vivo* activity, including interactions with protein micro-environments and other cell components [3].

FORCES CONTRIBUTING TO PROTEIN–NP INTERACTIONS

Van der Waals Interactions

When the outer electron clouds of two atoms get close enough to contact, an attraction develops. It is vital to remember that these interactions are distance-dependent, decreasing by the sixth power of distance. There are around four interactions every second, each with an average kinetic energy of about four-thousandths of one-millionth of the average kinetic energy of an atom in solution. When many contacts are connected (as in interactions of complementary surfaces), kinetic energy is more important. Under ideal conditions, Van der Waal interactions may provide bonding energies of up to 40 kJ mol. Atoms are unable to live in close proximity to one another because of their closeness. Consequently, sterically conflicting surface groups might limit a molecule's ability to bind to another, resulting in high energy expenditure. Since this van der Waals repulsion inhibits many nonspecific contacts, it is more important than the favorable bonds discussed before to determine macromolecular interaction selectivity [4].

Van der Waals forces, which range from 0.5 to 1 kcal/mol for nonionic van der Waals interactions, are modest. Oxygen and nitrogen are neutral molecules with electronegative atoms that prefer to attract electrons from their less electronegative neighbor atoms through the covalent connection.

Dispersion of charge ($\delta 1$) and ($\delta 2$) is created by this process, which introduces dipolarity into the molecule, generating charge dispersion throughout the molecule. By matching the positive and negative ends of one molecule to the other, it is possible to establish a weak contact force between two molecules. The mechanism of charge dispersion in participating molecules and the link generated as a consequence allowed us to classify these forces. Electronegative solid atoms in the molecules themselves allow them to generate a dipole because they allow for the dispersion of partial charges. Dipoles may be produced in a neighbor molecule by using permanent dipole agents. The interplay between the two is feasible. Suppose a link is created between two such permanent dipoles. The contact takes on an ionic aspect and is substantially weaker in such a situation. The force in issue is known as a Keesom force in this case. Consider what happens when an electrostatic dipole causes its following molecule's electron cloud to become dipolar and establish a connection with the dipole. The two molecules may interact in this situation. Debye forces are the term used to describe the force in this case.

In addition to the London force, a similar force, it provides partial charge dispersion between two neutral species that are adjacent to one another (*e.g.,* aliphatic hydrocarbon). Under normal conditions, the London force maintains the fluidity and cohesiveness of biological membranes. Polarity is generated in each other by nonpolar molecules. To put it another way, the London force is a product of this dipole-induced dipole interaction. The Keesom force is the name given to the weak attraction between dipoles. The Debye force, on the other hand, refers to a dipole-induced dipole interaction. In all three cases, the potential energy of the van der Waals force is inversely related to the separation distance.

H-bonds

Covalently connected donor H atoms with a partial positive charge, Δ+ (resulting from electron removal by a covalently linked O or N), form hydrogen bonds. With acceptor H atoms with a partial negative charge, Δ- (typically O or N).

H atoms are aimed straight towards the acceptor atom in these bonds, with the maximum bond strength (12 to 29 kJ mol). Protein-helices and -sheets and DNA and RNA base pairing depend on hydrogen bonding to stabilize their secondary structures [4].

Electrostatic Interactions

Electrostatic bonds are formed by charged groups that have gained or lost a proton. It has been postulated that hydrogen bonds, which are as strong as 20 kJ/mol, do not affect the structure of biological systems. To balance a positively charged ion, an inorganic counterion (such as Na^+ or Cl^-) is usually surrounded by water molecules. There are so many water molecules floating about that it is impossible to know exactly where the counterion is in relation to the charged group. As charges on potential macromolecule binding surfaces reject one another, electrostatic interactions are important in dissolving biological formations. To dissociate biological structures, electrostatic interactions are critical because the charges on each possible macromolecule binding surface are in direct opposition. The enormous spectrum of biological interactions that phosphorylation regulates is made possible. If an enzyme's phosphate group is added, it can destabilize a protein-protein interaction. However, the interaction may persist after the phosphate group is removed [4].

Hydrophobic Interactions

A rise in ΔS may be seen as a result of laboratory experiments. From the thermodynamics perspective, self-assembly and other association events involving combining distinct molecules to create more ordered structures may seem implausible. Despite this, a large number of binding reactions are very well-favored. Is it possible to explain how the interaction of molecules might lead to increased disorder? The entropy of the macromolecules and the solvent as a whole rise as the water around the macromolecules lose its order. A rise in water entropy partially offsets entropy gains from connected macromolecules. Hydroxyl bonds are the glue that holds bulk water together, making it a semi-structured solvent. Lipids and proteins' nonpolar (hydrophobic) portions cannot establish hydrogen bonds with water.

On the contrary, water molecules form "clathrates" or "cages" along hydrophobic surfaces that have a high concentration of water molecules bound by hydrogen bonds. Structured clathrates are more organized than aqueous water and water that interacts with polar amino acids. Hydrophobic groups are tucked away in recesses or between water-repellent surfaces as proteins fold, forming bilayers with macromolecules and phospholipids. By dispersing into the bulk phase, the formerly highly ordered water attached to these surfaces has increased the system's entropy. When macromolecules' hydrophobic areas are submerged, water disorder increases. The molecule surfaces must be complementary to prevent water from facilitating a specific intermolecular contact through hydrophobic interactions [4].

Salt Bridge

Amino acid salt bridges are formed when amino acids with polar opposing charges come together. A hydrogen bond may be formed between at least two heavy atoms. Water is the most common solvent to interact with these proteins' solvent-exposed regions. Polar water molecules shield coulomb interactions between opposing charges. A contact ion pair or ionic group dissolved in a solvent might be the result of the process. A few thousand pounds per square inch is a common estimate for salt bridges' relative strength. The Lysine-Glutamine salt bridge was investigated in a vacuum and water to imitate a hydrophobic environment in the protein core versus a solvent-accessible area of the protein. The peptide was revealed to be molecular in a hydrophobic environment. It transforms into a zwitterionic state when water molecules are added and form a salt bridge. Water and various co-solvents and ions are necessary for proteins to function in a live cell. By sheltering charges and competing to generate amino-acid–free ion couples, free ions in solution may affect salt bridges (specific influence). Further hydration degrades salt bridges to the point that they are no longer functional.

To better understand how proteins interact with dissolved salts, scientists have been attempting to address the questions of what occurs in solution to big and tiny salt ions, as well as ion pairing. More than any other aspect of the Hofmeister ion series, salting in/salting out is the most common motif in the literature. Water's ability to interact with protein surfaces was thought to be altered when ions were present, which altered the water's structure. There are two types of ion names for hydrogen bond network rigidity and flexibility, which are called "kosmotropic" and "chaotropic," respectively (chaotropic). If solvating water molecules or protein surface features ('enslaved water,' or "enslaving water") drive protein behavior, this is a fascinating subject to consider. These ion-specific effects on water structure have been proved to extend beyond the initial water molecule in direct contact with ions, both experimentally (through NMR measurements) and theoretically (at various levels of theory). Specific ions and their concentrations have a substantial effect on water dynamics; according to the authors, Ion-specific effects are evident at low salt concentrations, but the impact at high salt concentrations is unspecific due to increased viscosity – slowing water dynamics – making it difficult to distinguish between ion-specific and non-ion-specific effects [5].

THERMODYNAMICS OF PROTEIN–NP INTERACTION

There is an enthalpy gain ($\Delta H > 0$, which is somewhat offset by an unfavorable loss of entropy) in ChT complexation with all particles ($\Delta S > 0$, which results in free energy shifts of -32.2 to 34.4 kJ mol^{-1}). The complexation of histone and CytC NPs has an endothermic effect on the enthalpy contributions to the free energy of association ($\Delta H > 0$). A high favorable entropy change ($\Delta S > 0$) dominates the interaction of histone and CytC. Non-covalent forces such as electrostatic, hydrophobic, hydrogen bonding, and -stacking, and the desolvation of NPs and proteins all play important roles in the complexation behavior of proteins. Using eq 4-1, it is possible to simplify the complexation procedure:

$$protein.mH_2O + NP.pH_2O \leftrightarrow [protein.NP].(m + p.x)H_2O + xH_2O \qquad (1)$$

Non-covalent bond formation and solvent rearrangement coincide in the thermodynamics of complexation, according to eq 1. The non-covalent bond formation is exothermic from an enthalpic perspective ($\Delta H_{intrinsic} < 0$). In addition, the breakup of well-defined solvent shells is endothermic in the presence of heat. ($\Delta H_{desolv} > 0$). According to the observed negative enthalpy changes, establishing intrinsic bonds (or protein-particle interaction) is the most important factor in the complex formation of ChT with NPs. According to one theory, solvent rearrangement may play a role in enthalpy changes during protein-ligand interactions. Because the solvent-accessible surface area is often reduced, highly ordered molecules are discharged into the bulk solution due to the complex formation. Due to unfavorable desolvation and favorable intrinsic enthalpy, observable enthalpy changes are compensatory.

Interaction surfaces may benefit from water molecules at interfaces, but this does not mean that the complex interface maintains or increases hydration relative to free proteins and/or proteins and particles. The entropy change may be influenced by the conformational constraint of amino acid residues in the two partners.

ChT-NP complexation results in adverse entropy changes when desolvation does not compensate for the loss of entropy owing to the decrease in solute freedom. Unmistakable evidence of molecular disarray is provided by the huge increase in positive entropy during the complexation of NPs with histone and CytC. Histone and CytC have more charged residues than ChT. This is thought to be caused by the binding interface, releasing a significant quantity of water during complex formation.

As a result, there are many more polar surfaces on the associated interaction interfaces. NP-histone and NP-CytC interaction entropies will likely grow when more water is released from hydration or as water molecules are dissociated from a more ordered starting state.

A small amount of entropy contributes to the stability of a complex, but this is countered by unfavorable enthalpy shifts when well-defined solvent-protein and/or solvent-NP bonds are broken [6].

PROTEIN ADSORPTION TO NPS AS DESCRIBED BY CLASSICAL LIGAND-BINDING MODELS DEVELOPED IN BIOCHEMISTRY

As mentioned, protein adsorption onto NPs is physicochemically similar to ligand binding. After binding, desorbing proteins may remove additional proteins, and even the ligand shell from the NPs' surrounds. Simulating a solution of inert NPs in contact with a single protein, P., will be our first step. In terms of the reaction equation, let us investigate protein adsorption:

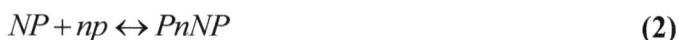

$$NP + np \leftrightarrow PnNP \qquad (2)$$

Proteins may bind to each NP's n binding sites in this manner; hence PnNP possesses n proteins. According to the law of mass action, a hemoglobin oxygenation forecast should not be based on an "all-or-none" approach that excludes partially saturated species. The dissociation equilibrium coefficient, or K_D, is the ratio of the off-to-on protein binding rates to NPs in thermal equilibrium.

$$K_D = \frac{c(NP)c^n(P)}{c(PnNP)} = \frac{K_{off}}{K_{on}} \tag{3}$$

For each NP, unbound protein, and protein–NP complex, the amounts of c (NP), c (P), and CnNP are measured. The concentration at which half of the NPs are saturated with proteins, K_{D0}, replaces the dissociation coefficient, K_D. This means that the concentration of free NPs and proteins in two separate solutions will be c (NP) and c(P), respectively when equilibrium is reached. Protein–NP complexes will be present in the solution with concentration c(PnNP) [7].

$$k'_D = k_D^{1/n} \tag{4}$$

SURFACE PROPERTIES OF NPS

Effect of Charge

As far as protein corona formation and subsequent biodistribution are concerned, NPs' surface charge is crucial. Examples include the rapid recognition of some positive NPs, which leads to their elimination from the body and ultimately to their removal by the RES and mononuclear phagocytic systems, which results in a reduced application yield. Functionalization with negatively charged groups (carboxylated, sulfate, *etc.*) may stabilize many NPs in physiological buffers with a negative ζ potential of 30-50 mV. A surface charge of 5 to 10 mV negative is lowered when these NPs come into contact with biological fluids like blood. For the colloidal particles to be stable, an electrostatic action alone cannot account for their behavior. In contrast, the protein corona's complexity is closely linked. NPs opsonization may be minimized by adding PEG to the NP surface to prevent nonspecific binding and enable particular surface groups to attach to receptors on cell surfaces.

Effects of Smoothness/Roughness

Nanoscale surface roughness may also change surface effects. It has been hypothesized that the roughness of NPs' surfaces may help them adhere to synthetic membranes, leading to better cell encapsulation. Nanoparticle-cell interactions are weaker when the corona shell composition of NPs varies by a small radius, which is attributed to the surface asperities.

Electron Transfer Capability

When used as electrochromic materials, catalysts, or pigments, TiO_2 NPs and SiO_2 NPs can conduct electrons. They may also be used as food additives as, sunscreen and cosmetic ingredients. Electron confinement or the formation of electron-hole pairs may result in ROS due to damage to biomolecules by these NPs. Tiny 63nm TiO_2 NPs were tested on human lung epithelial cell line A549 *in vitro*. The Comet test measured the extent of DNA damage and oxidative lesions. Fluorescence-sensitive fluoroprobe 20,70-dichlorofluorescein diacetate (DCFH-DA) was used to assess ROS generation in the cells. DNA damage was substantial, whereas DCFH-DA identified only a small amount of ROS.

When the TiO_2 NPs toxic impact is scaled from 63 to 200 nm, the protein and Ca^{2+} ion absorption on the TiO_2 NPs surface is enhanced, resulting in more induced oxidative damage. According to these and other studies, semiconductor NPs produce ROS, deplete glutathione, and experience harmful oxidative stress due to electron-hole pair formation, photo activity, and redox. TiO_2 NP surfaces may be coated or capped with a variety of molecules to decrease protein conformational changes during adsorption. These molecules include surfactants, polymers, complexing ligands, low-molecular-weight antioxidants, enzyme scavengers, *etc.* New antigenic sites (cryptic epitopes) may form due to protein denature on an NP surface, resulting in an immune response and autoimmune illness.

Effects of Hydrophobicity/Hydrophilicity

For example, proteins have been discovered to have a strong affinity for the surface of NPs, and this affinity has been linked to the level of structural alteration. Hydrophobic surfaces have a higher affinity for adsorbed proteins than hydrophilic surfaces, leading to severe protein denaturation if the proteins are adsorbed on hydrophobic surfaces. The isoelectric points of proteins may also influence how NP surfaces adsorb them. An increase in electrostatic interaction has also been linked to a decrease in the amount of structural change. However, despite the fact that hydrophobic contacts seem to predominate in the majority of cases, electrostatic interactions cannot be ignored, as shown by published research. As particle hydrophobicity grows, so does the number of protein molecules bound (stoichiometry), as shown by Cedervall *et al.* using ITC. Proteins have different affinities for polymeric particles with different degrees of hydrophobicity.

Protein/NP Ratio

There is a correlation between NP concentration and binding properties. This ratio is determined by the number of proteins binding to the particles and their density.

Reduced protein-to-NP density has been shown to boost binding and conformational changes between proteins because of greater energy transfer from the NPs to the individual protein molecules, according to experts. Polystyrene NPs with various surface charges and diameters of 50 and 100 nm were found to have the optimum protein solution to NP surface ratio, which was found to be 2.8 mL/m^2. Using centrifugation, separating the NP-protein coronas from the unattached proteins is required for further analysis. SPION-transferrin interactions were studied by Mahmoudi *et al.* using an ideal protein-NP surface ratio. SPION concentrations were raised 10- and 100-fold to test the r effect, which increased transferrin conformational relaxation upon adsorption. Smaller NP-protein interactions, on the other hand, make it more difficult to separate proteins from NPs during the magnetic separation process. The authors used electrophoreses based on sodium dodecyl sulfate and polyacrylamide gels. Reduced protein-SPION binding energy resulted in a substantial increase in protein concentration on SPION surfaces. Compared to SPIONs coated with poly (vinyl alcohol), the protein's SDS-PAGE bandwidth was lowered. The protein corona, binding affinity, and the association/dissociation constants of NPs and their concentration were shown to be strongly correlated in this investigation. Even yet, it is important to consider how these effects hold up in circumstances where there is competition for binding resources. The corona's composition may also be influenced by variations in the protein-to-NP surface ratio. Proteins like HSA and fibrinogen, for example, may briefly coat the surface of NPs. They may eventually be overtaken by less abundant but more specific protein pairs that are more sensitive and slower to react. If r lowers to an undesirable level, a lower affinity binding protein, such as albumin, may take over the NP surface.

Effect of Functional Groups and Targeting Moieties

Endothelium cells were used to replicate the movement of polystyrene NPs through the circulatory system, according to Ehrenberg and colleagues. Protein-adsorbing properties of the different NP surfaces suggest their potential to interface with cells. Consequently, it was deduced that NP-cell interaction was independent of identifying the protein. NP protein adsorption was shown to be fastest in the seconds to minutes range, suggesting that proteins on the surface of NPs may play an important role in cell interaction over considerably longer time scales [2].

FACTORS AND EFFECTS

Factors Affecting Protein-nanoparticle Interactions

Attaching certain chemical molecules may alter properties like charge, hydrophobicity, and surface chemical characteristics. The physicochemical properties of the core material may be altered by adding a surface modification to the NPs.

The NPs' biocompatibility and biodegradability are determined by their core composition and surface characteristics, impacting NP effectiveness in any application. The charge of NPs plays a significant role in forming NP–protein coronas and the *in vivo* biodistribution of NPs by influencing their interactions with proteins. Aubin-Tam investigated the influence of NP surface charge and Hamad-Schifferli utilizing AuNPs functionalized with positive, negative, or neutral (polyethylene glycol) thiol (PEG) ligands linked to particular protein locations, such as C102 of Saccharomyces cerevisiae Cyt C (C102). A significant effect on protein structure was seen when functionalized ligands were used to specify charge on the NP surface. At C102, NP ligands and amino acids interact to influence protein stability, resulting in protein denaturation.

Sasidharan *et al.* have shown that surface charge affects NP–protein interactions in new ways. Copper and lipoic acid-coated silver NPs were used to evaluate the kinetics of adsorption, individual protein coronal development, and the effect of this creation on the dispersion and stability of these proteins. Positive coincubation with IgG shifted from a negative to a positive charge on the NP surface of lipoic acid–NPs and citrate–NPs. Lipoic acid–NP–IgG coronas had greater zeta-potential values than citrate–NP–IgG coronas, which indicates that lipoic acid–NP–IgG coronas are more stable. The citrate, on the other hand, is much less compatible with NPs.

After positive coincubation with IgG, the net charge of the NP surface in lipoic acid–NPs and citrate–NPs changed from a negative to a positive. This suggests that the lipoic acid–NP–IgG coronas are more stable, as shown by the higher zeta-potential values than those for citrate-NP–IgG. On the other hand, Citrate has a much lower affinity for NP surfaces. Proteins' high affinity for hydrophobic surfaces, relative to proteins' structure when adsorbing on hydrophilic surfaces, may cause significant denaturation of the NP surface. As NP surfaces became more hydrophobic, the quantity of adsorbed protein on them increased, as shown by Cedervall *et al.*

Human serum proteins were tested for their ability to bind to SNA–NP conjugates of spherical nucleic acids. SNAs vary from linear nucleic acids because they are composed of spherical NPs functionalized with highly orientated oligonucleotides. The creation of protein corona was revealed to be regulated by oligonucleotide-modified NP structures. Due to the rapid uptake of SNAs, this design may be employed in various therapeutic and diagnostic applications. Oligonucleotides are more stable and resistant to deterioration than linear forms because of their thick layer. Genomic DNA sequences, especially G-rich sequences, have been shown to promote connections between proteins that bind to SNAs and cell surface scavengers.

The first time this was found, G-rich sequences and 13 nm AuNPs were used to generate SNAs that could be used to study the influence of these G-quadruplex structures on proteins exposed to serum. As a check, the researchers used poly-T sequences.

At 37°C for 24 hours, SNAs were grown in 10% human serum to determine their binding ability to proteins. Poly-T SNAs were smaller than G-rich SNAs. More serum proteins are drawn to SNAs that are G-rich. More protein-protein interaction clusters are seen on G-rich SNAs than on other SNAs. As a consequence, 82 proteins were recovered from G-rich SNAs, and 54 proteins were collected from poly-T SNAs, with 49 common proteins and G-rich SNAs adsorbing more proteins than poly-T SNAs. According to the findings, g-rich SNAs adsorb more human blood proteins, including those from the immune system, than poly-T SNAs. Both SNAs have different amounts and compositions of proteins on their surfaces. As shown in this work, the DNA sequence and the DNA structure of an NP had different roles in the protein binding capabilities [8].

NPS AS INHIBITORS OF PEPTIDE AND PROTEIN AGGREGATION

Fibril production is restricted by the primary approaches used to address neurodegenerative diseases. Oligomers and monomers that bind differently can be used to stop this from happening. Also, oligomers are one of the A agglomerates' most neurotoxic species, so blocking them with NPs will be very bad from a biological point of view. Nanoparticle interaction with proteins during fibrillogenesis is being studied in many ways. When thioflavin-T interacts with amyloid fibrils, the emission spectra shift toward red. Soluble proteins, oligomers, and amorphous aggregates do not exhibit a comparable response. The use of circular dichroism and photoluminescence as additional tools for studying fibrillation control is critical. These approaches may distinguish unfolded regions of proteins using these approaches, revealing a kinetic shift toward these regions. For the first time, a crystallographic model by Heldt *et al.* showed that insulin amyloid fibril formation is often unidirectional. Focusing on the factors that cause fibrillation is essential for designing an effective inhibitor and a nanomolecular species that can be used for this. Fibril creation relies heavily on interactions like van der Waals and hydrogen bonds.

There are molecular conformational features that are necessary for inhibitors in order to attach to nuclei or other oligomers in order to minimize their harmful effects in a three-stage kinetic model of amyloid formation developed by Lee *et al.* By altering the free monomeric peptides available, NPs may minimize fibrillation, yet fibril formation is inevitable. NPs can interfere with the nucleation phase, the polymerization phase, or even the equilibrium point (diverting the peptide from the polymerization pathway). Findings from this research show that the fibrillation process ceases in the lag phase when NPs are administered early in the procedure of the experiment. After the control lag time, NPs had no inhibitory effect.

Adding NPs does not influence monomer/oligomer interactions with NPs, suggesting that the elongation process is favorable after critical nuclei have been formed. If the NPs concentration is high, the fibrillation process will likely be slowed down because of the adsorption of monomeric fibrils or newly-produced polymers onto the NPs. If NPs are introduced after fibril production has started, the process may be reversed, or the fibrils may be destroyed. However, high protein concentrations may also limit fibril growth by preventing the production of fibrils owing to interactions between proteins and NPs.

NPs based on hydrophobic amino acids, such as those created by Skaat *et al.*, have been shown to influence the production of fibrils by accelerating the creation of oligomers and nucleation before fibrils are formed, thereby enhancing the fibril's ability to resist deformation. Suppose the solution's low concentration of monomers and oligomers causes nuclei disruption and interferes with elongation. Hydrophobic interactions in these NPs, such as pairs of FF residues, may be responsible for this phenomenon. A40 pre-fibril aggregates are well-liked by these interactions. Analyzing the structural changes in the presence of NPs (dendrimers), Milowska colleagues; employed circular dichroism to examine the effects of PAMAM dendrimers on -synuclein. A decrease in the -sheet structure was indicated by the emergence of a positive signal between 195 and 206 nm after incubation with the NPs. Due to their size and concentration, they may inhibit thermodynamic and kinetic reactions from dendrimer NPs. Kinetic inhibitors did not affect the final number of fibrils, independent of the lag time. Thermal inhibitors did not affect the pace of amyloid production, but they lowered the ultimate number of fibrils. The elongation rate of A aggregation is influenced by lower generation (G3) PAMAM dendrimers. This effect is less noticeable in higher generations (G4, G5).

According to Rekas *et al.*, during fibrillation, it was shown that PAMAM dendrimers affected -synuclein fibrillation and breakdown in preformed fibrils; dendrimers hinder the formation of -sheet structures by breaking down existing -

sheets or their agglomerates, according to the study team's findings. Milowska and colleagues have observed that dendrimers containing phosphorus may effectively quench tyrosine fluorescence since intrinsic tyrosine fluorescence intensity diminishes when the dendrimer/protein ratio is high. This mechanism intermolecularly links the dendrimer's tyrosine hydroxyl and cationic end groups. Also, phosphorus dendrimers' cationic groups may bind to -synuclein's basic amino acid N-terminal region and hinder fibril formation, as previously reported.

Dendrimers with a low concentration are more effective than dendrimers with a high concentration. They are also social, so they stay together and produce bigger clusters of algae (dimers or oligomers). Amyloid-peptide aggregates were significantly inhibited by sulfonated dendrimers, and the -sheet conformation was similarly decreased. Protein and peptide binding is made possible by saccharides on the surfaces of these sulfonated dendrimers. Electrostatic interaction with basic amino acid residues allows interaction with A peptides. According to Kim and Lee, attaching NPs to the hydrophobic motif may also inhibit. The authors characterized the key target for aggregation suppression, the KLVFF hydrophobic sections of Aβ peptide, as being bound by fullerene. It is possible to develop type 2 diabetes when non-native insulin aggregates and new insulin fibrils are created.

PHFBA (poly(2,2,3,3,4,4-heptafluorobutyl acrylate) NPs strongly suppressed insulin fibrils in the presence of Skaat and colleagues. Insulin fibril production is hindered by these NPs, preventing the transition from α-helix to β-sheets [9].

HUMAN SERUM ALBUMIN (HSA) PROTEIN NATURE

There are 585 amino acid residues in HSA's molecular mass of 67 kDa. Domenici *et al.* were the first to establish a connection between HSA-based CSPs and diol-silica particles. Mallik *et al.* recently used GMA and ethylene glycol dimethacrylate to attach HSA to polymer monoliths (EDMA). To synthesize aldehyde groups, periodic acid was used to oxidize diol-silica and polymer monoliths made of GMA and EDMA polymers, and a Shiff base was formed by reacting an amino group from HSA with the Shiff base, followed by reduction. Increased retention and greater or equivalent resolution or efficiency were seen for silica particles immobilized with HSA in monoliths or GMA/EDMA monoliths. A silica monolith might be used instead of GMA/EDMA monoliths as a CSP with immobilized HSA.

The SMCC and SIA procedures generated CSPs with equivalent or greater enantioselectivity and stability than those prepared using Schiff-based HSA for CSPs based on HSA-based methods. A wide range of 2-aryl propionic acid derivatives, such as temazepam, temazepam lorazepam, temazepam fenoprofen and temazepam lorazepam derivatives, and other benzodiazepines are resolved using HSA-based CSPs. Reduced folates like leucovorin and 5-methyltetrahydrofolate are also HSA Proteins discovered in HSA, and other mammalian species have a tight relationship. HSA and BSA have equal stereoselective binding properties. The elution sequence is inverted in CSPs based on HSA and BSA, with (S)-Wf eluting before (R)-Wf. For CSPs based on HSA, the elution order is the opposite of this. These results are in line with the enantioselectivity of natural proteins.

Tofisopam, a 2,3-benzodiazepine derivative, exhibits stereoselective binding in humans. However, in all other species, it exhibits the reverse (bovine, dog, horse, pig, rabbit, and rat). Dog albumin, on the other hand, has 1,4-benzodiazepine binding properties equivalent to those of HSA [10].

METHODS USED FOR INVESTIGATION OF THE INTERACTION OF PROTEINS WITH NPS PRODUCED BY THE GREEN SYNTHESIS METHOD

Dynamic Light Scattering (DLS)

The optical method of dynamic light scattering (DLS) may characterize scattered systems. It uses high-frequency light scattering to quantify microstructural process dynamics such as elastic vibrations in gel, the transition from sol to gel, and particle aggregation. The Brownian motion of individual particles in aqueous solutions is often measured using DLS for particle size analysis. Particles' hydrodynamic equivalent diameters may be determined by analyzing their intensity-weighted distribution (or hydrodynamic diameter). Analytical centrifugation and ultrasonic spectroscopy, which employ optical and hydrodynamic models, may be converted to a number- and volume-weighted size distribution. When calculating the hydrodynamic diameter, it is necessary to include individual particles and particle aggregates or agglomerates. In aggregates or agglomerations, the size of the individual particles does not have much of an impact on this feature.

It is possible to estimate the size distribution of representative samples using DLS, which uses the numerical modification of their spectra. This is the method of choice for particle size analysis in the sub-micrometer range. Therefore, it has a poorer resolution than counting or fractionating approaches when it comes to small details of size distribution. With DLS, you may combine fractionation and size analysis in a hyphenated measurement configuration. There are several practical and economic advantages to this, including speedy analysis and minimal expenses per measurement. Finally, new hardware and data processing improvements are improving the technique's capacity to deal with polydisperse materials and clarify fine features in the size distribution. Coherent light from scattering objects like large molecules or microscopic particles may be monitored over time using dynamic lighting scattering (DLS).

The variation of the measurement signals is analyzed in terms of their causes, which might come from a variety of sources. Thermal motion of the scattering items causes short-term (microseconds and even nanoseconds) fluctuations in DLS. Colloidal suspensions may be studied using DLS to determine phase transitions and the elastic properties of the gel they contain. Submicrometer particles are often measured using DLS. With this method, a fluctuating signal is produced at the detector because it takes advantage of the particles' Brownian motion, which permanently rearranges space and alters the interference between scattered signals. Kinetics of rearrangement are dictated by the particle diffusion coefficient Dp, which is inversely related to particle size:

$$D_p = \frac{K_B T}{3\pi\eta\chi_{h,t}}$$

(5)

The Boltzmann constant (kB) and temperature (T) are the two variables in the Stokes-Einstein equation. The fluid's dynamic viscosity (η) is given by $X_{h,t}$, which gives the translational motion's hydrodynamic diameter. A particle's random orientation determines its equivalent diameter, which might vary greatly from other equivalent diameters. Aggregates with $X_{h,t}$ less than the diameter of the sphere but much larger than Stokes or even the component particles themselves are especially vulnerable to these changes.

Assuming Brownian motion is to blame for signal variations, various micro-processes such as sedimentation, agglomeration, or multiple scattering that change the spatial configuration or degrade the coherence of the light signal are presupposed when assuming Brownian motion is to blame. Particles less than "10 nm" may only be used in extremely dilute, well-stabilized suspensions or emulsions for DLS particle size characterization. NPs with low polydispersity may be used in a number of ways, such as carriers for functionalized chemicals or as UV-absorbing additives.

DLS is a typical measuring method for determining the characteristics of these NPs. DLS has remained popular for this kind of analysis throughout the years because of the ease of use, fast measurement periods, and cheap running costs of the equipment. Many interlaboratory comparisons have also shown the validity of DLS. They all agree that the method has pretty high repeatability and strong reproducibility for samples that have been stabilized.

First-generation DLS equipment was used to investigate DLS studies of monomodal and bimodal latices in 200–800 nanometers. Polydispersity, however, had far higher variability than the 3 percent they used to quantify the uncertainty in the effective hydrodynamic diameter (X_{cum}). Colloidal silica performance in DLS was examined in more recent research (approx. 19 nm, low polydispersity). For a 95 percent confidence level, the total uncertainty in x cum was 3 percent. Volume-weighted distributions, on the other hand, had a substantially larger mean size than volume-weighted distributions, notwithstanding this fact. This was confirmed in a second study that used a different chemical for testing. On the other hand, fractal-like aggregates with modest polydispersity were evaluated in an interlaboratory comparison on DLS. According to the data, the effective hydrodynamic diameter (X_{cum}) varies across 10 labs by less than one percent.

Numerous numerical representations of nanoparticle size distribution exist, including the number-weighted size distribution (q_0). The mathematical transformation into a number-weighted distribution of sizes is simple to understand and implement. The intensity weights of the different size fractions influence correlation functions and frequency spectra, but this does not change the main DLS conclusions. So DLS inherently provides size distributions with intensity weighting (Q_{int}). However, near the fine edge of the distribution function, it magnifies any noise or artifact result.

Distribution analysis in instrument software often yields a wide range of number-weighted size distributions. This is especially true for NPs sensitive to optical Rayleigh scattering. Many researchers have devised a different method for calculating number-weighted size distributions (q_0). Two-parametric functions, such as the gamma or log-normal, may be used to approximate the size distribution. It is possible to utilize particle size k to express cumulant analysis results like xcum and the polydispersity index. In this way, the analytical distribution functions can be located. The polydispersity parameter for the distribution's polydispersity may be accurately determined (Q_{int}). Rayleigh scattering for NPs may be utilized to convert Q_{int} to q_0. For low and intermediate polydispersity, the method worked well. For example, it may be applied to non-sphere-like particles such as vesicles by relying on Rayleigh–Debye–Gans scattering [11].

Zeta Potential Measurements

If a positive unit charge were sent from infinity to the surface without speeding up, the electric potential of the surface would be equal to the amount of work it would take to do so. Let us say an electric field is put on a moving colloidal particle. In this case, ZP is the electrokinetic potential (EP), which is the potential at the slipping/shearing plane.

The EDL (electric double layer) and the dispersion layer surrounding electrophoretically mobile particles have a potential difference at the sliding plane reflected in the ZP.

Understanding the EDL and Slipping Plane

Adsorbed double layers, also known as EDLs, form on the surface of scattered charged particles. The inner layer is dominated by ions/molecules with the opposite charge of the particles.

Debye's law says that for every Debye length beyond the Stern layer, the field weakens by a factor of $1/e$. This means that the surface charge has less of an effect on the particles' electric fields. Only a few nanometers beyond the particle's surface may be seen in this electrostatic effect, although this effect should theoretically stretch infinite. The EDL results from NPs' electrostatic fields creating a diffuse layer of the same and the oppositely charged ions/molecules. Several parameters, including pH, ionic strength, and concentration, determine the makeup of this diffuse layer.

The dispersion of charged particles toward the opposing electrode is caused by an electric field (electrophoresis). Electrophoresis takes place inside a diffuse layer. A hypothetical plane serves as an interface in the dispersion layer around the moving particles. The particle-fluid interface potential is represented by ZP in the slipping/shear plane. The Greek letter (ζ) was initially employed in mathematical equations to denote the phenomena known as "zeta potential". The Nernst potential ($\psi 0$) refers to the surface potential of the particle and cannot be measured. With increasing distance from the particle surface, the electrostatic field weakens.

$$\psi = \psi d e^{-kx} \tag{6}$$

Where ψ = surface potential at a distance x from the stern layer, ψd = surface potential at the stern layer, κ = Debye-Hückel parameter, x = distance.

When the slipping plane is close to the stern layer – the $\psi d \approx \zeta$ and hence, Eq. (8) can be modified as an equation:

$$\psi = \zeta e^{-kx} \tag{7}$$

When the ionic strength is high, the Debye-Hückel parameter (κ) is smaller. As a result, the ZP drops as the double layer is squashed by the increasing ionic presence.

Fundamental Mathematical Operators While Measuring ZP

To calculate ZP, we must use a method that does not directly assess the electrophoretic mobility of charged particles in an applied electric field. The particles' electrophoretic mobility (e) is first computed as (Eq. (8)):

$$\mu_e = \frac{V}{E} \tag{8}$$

V is the particle's speed in (μm/s), and E is the electric field intensity in (V/cm). Henry's equation (Eq. (9)) is used to compute the ZP from the acquired μe.

$$\mu_e = \frac{2\varepsilon_r \varepsilon_0 \zeta f(ka)}{3\eta} \tag{9}$$

Relative permittivity/dielectric constant (εr), vacuum permittivity ($\varepsilon 0$), ζ (ZP), Henry's function f(Ka), and temperature-dependent viscosity (η) are all defined. f(Ka) is considered to be 1.5 when the EDL thickness is much less than the particle radius, which may be due to bigger particles (up to 1 μm) in high salt concentration (10^{-2} M) aqueous solutions. Henry's equation then modifies into the Helmholtz-Smoluchowski (HS) equation (Eq. (10)):

$$\mu_e = \frac{\varepsilon_r \varepsilon_0 \zeta 3\eta}{\eta} \tag{10}$$

Nano-DDS development relies heavily on the HS equation, which applies to a wide range of medicinal products. f(Ka) is regarded to be 1 when the EDL thickness is much bigger than the particle itself because of very small (less than 100 nm) particles scattered in very low salt concentrations (10^{-5} M). Henry's equation can be modified as the Hückel equation (Eq. (11)):

$$\mu_e = \frac{2\varepsilon_r \varepsilon_0 \zeta}{3\eta} \tag{11}$$

Since aqueous dispersions do not adhere to the Hückel equation, pharmaceutical preparations are seldom affected. Ceramics, on the other hand, is popular [12].

Aromatic Amino Acids

Microbes primarily produce aromatic amino acids. Since the initial discovery of bottleneck enzymes in the biosynthetic pathway and their regulation, *i.e.,* (1) 3-deoxy-Darabinoheptulosonate 7-phosphate (DAHP) synthase that mediates stereo-specific condensation of erythrose 4-phosphate (E4P) and PEP to generate DAHP; (2) chorismate mutase/prephenate dehydratase or the chorismate mutase/prephenate dehydrogenase complex that catalyzes the conversion of chorismate to phenylpyruvate or 4-hydroxyphenylpyruvate *via* prephenate, and (3) transcriptional regulation of encoded pathway enzymes such as tyrR in E. coli that codes for a transcriptional regulator; the strategies to deregulate the feedback mechanisms and to modulate the transcriptional regulator activity had been employed and assessed in different organisms to improve the amino acid production. Several research has employed logical approaches to boost the production of certain aromatic amino acids. The results have been discussed extensively in a number of academic journals. As a reminder, here are some prominent examples. After 36 hours of incubation, recombinant E. coli with a rationally modified aromatic amino acid biosynthesis pathway produced L-tyrosine at 13.8 g/L. With proteomics and metabolic profiling, they could eliminate the alleged bottlenecks, YdiB responsible for the intermediate accumulation and the low expression of dehydroquinate synthase (AroB) required for shikimate formation, to produce an L-tyrosine-producing strain that achieved 80% of the theoretical maximum yield. Alternative C-sources for aromatic amino acid synthesis have been intensively researched. Short regulatory RNAs were used to combine the combinatorial knockdown of four possible genes in four unique E. coli strains, resulting in a titer of 0.37 g/L per hour of L-phenylalanine synthesis from glycerol.

Synbiotic approaches, including those using *C. glutamicum*, *E. coli*, and *S. cerevisiae*, have been used to attempt the production of aromatic amino acids and their derivatives from these diverse chassis organisms. Readers are urged to reference recently published articles for further information [13].

Intrinsic Fluorescence Measurements

Fluorescence is a technology that uses a small amount of material. When it comes to biochemists or molecular biologists, fluorescence is one of the most often employed tools they have at their disposal. In most fluorescence measurements, just a few nanomoles of analyte are needed, but in other experimental designs, only a single molecule is needed. Intrinsic fluorescence probes for proteins are tyrosine and tryptophan atoms. To a greater extent than tyrosine, tryptophan fluorescence predominates in proteins that include aromatic residues.

Adding or removing tryptophan residues from specified places in a protein has become virtually normal in recent molecular biology developments. As a consequence, the bulk of studies on intrinsic protein fluorescence concentrates on tryptophan residues. A protein's structure can only be detected in particular places owing to the limited tryptophan residues. Another option is to use covalently or non-covalently attached extrinsic fluorescence probes so that a range of fluorescence characteristics may be imparted to the protein; certain proteins already include intrinsic fluorophores.

Fluorescence is a useful tool for obtaining information about protein structures because of the environmental sensitivity of the fluorescence signal. Using tryptophan's sensitive indole side chain emission spectra, one may distinguish between native and unfolded protein states. Fluorescence emission competes with other molecular activities on the time scale of the emission process, resulting in this environmental sensitivity. At the same time as molecular rotation and translation and the movement of protein side chains, photon emission may occur.

Collisions with quenching groups or molecules may deactivate excited states in fluorophores, leading to detectable depolarization of emitted light by dipolar relaxation of polar groups and water around the excited state in the rotational motion of the fluorophores on the emission time scale. As long as the D-A distance is near enough, a donor (D) fluorophore may transmit resonance energy to an acceptor (A) on a time scale comparable to the emission process. The dipoles' orientation is not critical. When crystal or NMR structures are not viable for large multi-protein complexes, such data on energy transfer may be assessed to get the D$\rightarrow$A distance.

Fluorescence's environmental and emotional sensitivities may be seen experimentally since the approach is multi-dimensional. Excitation and emission wavelengths of fluorescence may be used to create spectra. It is possible to obtain fluorescence decay profiles by observing changes in fluorescence intensity over time. Quanter (or other agents such as protons or co-solvent) may measure the intensity and provide information on dynamic accessibility and other close interactions. As the polarizer angle changes, the fluorophore's rotating motion may be measured.

Additional measurements may be done, such as the anisotropy decay time-resolved or the anisotropy intensity *vs.* wavelength and quencher concentration (time-resolved anisotropies). Fluorescence's ability to operate in several dimensions is quite beneficial. It is not possible to resolve fluorophores like tryptophan in a protein along the wavelength axis. There are times when "Experimental Axes" of time, quencher concentration, and polarizer angle may be utilized to resolve them. Individual tryptophan spectra and decay times may be identified by combining these axes and/or studying mutant proteins with different numbers of tryptophan residues. Because of this, it is often hard to distinguish distinct spectra for individual tryptophan residues when there are three or more emission sites.

Fluorescence may be used in various instrument designs, another key benefit. A sample has to be illuminated from both the inside and the outside. Because of the need for short measurement durations in transient kinetics investigations, strong signal-to-noise ratios, as well as speed, are critical in the measurements. Capillaries, stopped-flow cells, high-pressure cells, and microscope slides are among the unique fluorescence measurement settings.

Protein kinetics and low-resolution structural information may be gleaned from fluorescence. When it comes to fluorescence, however, the most prevalent use is as a probe for conformational changes of proteins, such as protein unfolding transitions as well as ligand binding and protein-protein interaction activities. Thermodynamic and kinetic data may be gathered using these applications. The difference in fluorescence signals between various protein states makes these applications work. Thermodynamic and kinetic data may be collected if a change in fluorescence is independent of the reason. When it comes to thermodynamics and kinetic applications like these, fluorescence is an excellent choice because of its experimental advantages (broad concentration range, short measurement time, and flexible instrument configuration).

Fluorescence characteristics of proteins have been the subject of much research, including efforts to connect fluorescence lifetimes and anisotropy degradation rates with molecular dynamics models. The fluorescence qualities of a large number of proteins containing tryptophan, on the other hand, will be of more service to a novice user of this method [14].

Quenching Mechanism

When fluorescence intensity is reduced, the process of fluorescence quenching is called Quenching may happen from several molecular interactions. To name just a few: molecular rearrangements and collisional quenching are all examples of excited-state activities.

A fluorophore and a quencher clash to produce collisional or dynamic quenching, which is the subject of this chapter. We will discuss static quenching in order to understand better how the quencher binds to the fluorescent sample. Data analysis may be made more difficult by the use of static quenching. The sample's optical characteristics may cause apparent quenching in addition to the mechanisms discussed above. Fluorescence intensity may be reduced by high optical density or turbidity. The chemical information included in this insignificant kind of quenching is sparse. The declines in fluorescence intensity are assumed not to be the result of such insignificant effects throughout this chapter. Studies on fluorescence quenching have yielded valuable information about biological systems since it is a fundamental occurrence. Quenching has several biological applications because of the molecular interactions that lead to quenching. The fluorophore and quencher must interact molecularly for both static and dynamic quenching to work. Collisional quenching can only be achieved if the quencher diffuses continuously to the fluorophore. No photons are emitted when the fluorophore returns to the ground state after contact. In the absence of a photochemical reaction, quenching is the most prevalent fracture.

Static quenching results in the formation of a nonfluorescent compound between the quencher and the fluorophore. The fluorophore and quencher must be in direct physical contact for static or dynamic quenching to work. The necessity for molecular contact is at the root of many quenching applications. Fluorophores' accessibility to quenchers may be determined through quenching measurements. Quenchers and fluorophores may be incorporated into macromolecules impervious to the quencher. The quencher will be unable to penetrate the macromolecule in this situation. In this case, there is no possibility of quenching. Some of the fluorophores found in proteins and membranes may be determined by these tests.

They can also determine the permeability of quenchers for these fluorophores. Learning to a Ceasefire The diffusion coefficient of the quencher may also be affected by the rate of collisional quenching. The fluorophore is affected by collisional quenching, which extends the solution's volume and distance. The root-mean-square distance $\sqrt{\overline{\Delta x^2}}$ That a quencher can diffuse during the lifetime of the excited state (τ) is given by $\sqrt{\overline{\Delta x^2}} = \sqrt{2D\tau}$ where D is the diffusion coefficient. Consider an oxygen molecule in water at 25°C.

2.5×10^{-5} cm^2/s is the coefficient of diffusion for this substance. The oxygen molecule may move 45 during a normal fluorescence lifetime of 45 Å. Diffusion across much greater distances may be seen with a longer lifespan. If you have a lifespan of only twenty nanoseconds (ns), the average oxygen diffusion distance is 100 nanometers (nm). Diffusion over much greater distances may be detected using probes with microsecond lives that have a longer lifespan. The fluorescence quenching technique may thus be used to determine the diffusion of quenchers across quite long distances, equivalent to the size of proteins and membranes. Sorption relaxation is not applicable in this case. It is predominantly the solvent shell responsible for the fluorescence changes caused by the reorientation of the solvent molecules.

Thermodynamics of Protein-Ligand Association

The environment may greatly affect their function and structure since proteins are only somewhat stable. Water-soluble proteins are very sensitive to nonaqueous solvents, which reflects this. It is, at the very least, possible to explain the hydrophobic interactions in this case.

Solubility and direct calorimetric measurements were used to derive the thermodynamic parameters that define these interactions and figure out additivity relations to compute partial molecular heat capacity at infinite dilution of model substances such as hydrocarbons, alcohols, and amino acids.

For a wide variety of model compounds, the predicted and actual values are in perfect agreement, and the impacts of special groups, branching, and cyclization on the heat capacity of the molecules have been identified. Water's interaction with individual hydrophobic and hydrophilic groups has been studied in this research. However, their knowledge is far from sufficient to predict protein characteristics.

Refocusing attention on water's uniqueness as a solvent to which most proteins are exposed, however, has shifted the focus. Many experiments and theoretical studies have been carried out to discover what makes water tick in order to maximize protein activity in an aqueous solution structure. These subjects are covered in depth in volumes one through seven of this series. Water-protein interactions may be studied energetically using the heat capacity of proteins at various hydration levels. For example, this approach is separate from hydration studies, which only report the amount of nonfreezing water in the body. This number may be estimated using a variety of methods, including scan heat capacity studies, NMR investigations, and hydrodynamics. The results of heat capacity and NMR tests are often in accord with each other regarding water's ability to thaw. Protein solutions with water concentrations between 0.3 and 0.5 g per gram of macromolecule are not affected by bulk water properties. Hydrodynamic research estimations varied from the mean of 0.53 g H_2O per gram of protein for the 21 proteins studied. For the heat capacity-to-hydration curve, this comparison of the lysozyme-water system's infrared, diamagnetic susceptibility, and activity data yielded a precise chemical interpretation. This diagram depicts a particular series of occurrences. The overall specific heat capacity decreases linearly if the bulk water content is reduced to 0.38 grams per gram of protein. Hydrated protein's heat capacity plummets from its initial value of 4.18 J/g K to only 1.48 J/g K. One protein molecule is made up of just 300 water molecules, resulting in a water-t--protein ratio of 0.38 g H_2O/g.

It is time to provide lysozyme with a standard diluted-solution thermodynamic profile below 0.38 g. The overall heat capacity varies nonlinearly with weight fraction.

The nonthermodynamic measurements may be compared to the heat capacity *vs.* weight fraction graph, allowing the identification of three separate processes. The water-nonpolar interaction on the enzyme surface is thought to be the origin of the discrepancies in hydration levels of 0.38 and 0.25 g H_2O/g protein. For example, carboxylic and amide groups account for the difference between 0.25 and 0.07 grams of H_2O/g protein in heat capacity. The 0.05 g H_2O/g protein heat capacity limit may be due to proton redistribution activities that occur when the first water molecules come into contact with the protein. To put it another way, infrared absorption tests have shown that drying protein causes the pK values of carboxylic and basic groups to reverse. A protein's enzymatic activity starts before a water monolayer has formed. It is important to remember this. However, even after the thermodynamically measurable hydration changes, the specific activity continues to climb.

The protein may thus need many layers of H_2O in order to acquire its inherent kinetic properties. It was observed that there was an anomaly in the heat capacity-to-water-content curves for the enzyme X-chymotrypsin and the DNA from calves that were between 0.05- and 0.10-grams H_2O/gram of the enzyme, as assessed by vapor pressure measurements. These values are in agreement with the hydration entropy and enthalpy maximum. According to X-ray diffraction and dielectric and NMR relaxation studies, dehydration below this critical threshold produces structural alterations in the biopolymers that lead to extremes in thermodynamic excess functions. Recent studies also found statistically significant compensatory effects between excess enthalpy and entropy components.

Hydrophobic interactions have been known for a long time for maintaining the native protein structure. It is because of these studies that thermodynamic approaches (especially heat capacity measurements) may resolve mechanistic details when appropriately coupled with the findings of other techniques that this research has been widely discussed. A discussion over how much of a role hydrophobicity plays in polar and molecular interactions is ongoing.

Toluene, octane and octanes are different when measuring the enthalpy of phenol transfer into the water. According to this study, the lowest Gibbs energy of the three is toluene water, octane water, and toluene water. The following are the transfer entropies we may derive from this data: At 38.5 J/(molK), octane loses energy, but at 1.17 J/(molK), it gains energy (octane-octanol-water). Variations in nonaqueous medium interactions must account for differences in thermodynamic transfer characteristics since phenol should end up in the same ultimate state in water. As a result, the assumption that octanol and phenol have a strong hydrogen-bonded connection may explain the difference of 50 J/(mol K) in entropy between the two transfers of phenol from octane to water. The huge negative entropy changes and positive heat capacity increases when hydrophobic groups are exposed to water, regardless of the chemical cause, are extraordinary. The heat capacity of straight-chain alcohols, carboxylic acids, and N-substituted amides increased by around 84 J/(mole CH_2-group K) as the chain length grew, independent of the hydrophilic group type. This demonstrates that the increased heat capacity is related to a change in the solvent water's structure. A protein folding study further supports this view. The number of nonpolar group connections detected by X-ray analysis is often linked to variations in denaturation heat capacity. Hydrophobicity characteristics are used in Ikegami's hypothesis of protein unfolding, which yields the same findings.

Pure aqueous solvents provide a straightforward interpretation of model compounds and protein denaturation using thermodynamic data; mixed water-alcohol systems do not. As a consequence of their ability to alter the water structure around proteins and reduce hydrophobic interactions between nonpolar residues, alcohols make excellent model medicines. Isothermal flow microcalorimetry studies of transfer enthalpies at various temperatures have been integrated with differential scanning calorimetric findings extensively. Using the enthalpy of the lysozyme-aqueous propanol system, it is possible to apply numerous additivity relations. Aqueous I-propanol solutions interact with lysozyme, according to two recent studies. Adding I-propanol at first decreases the apparent specific heat but subsequently raises it dramatically. I-propanol denatures protein such that guanidine hydrochloride does not exist in an aqueous solution, which is the most exciting observation.

According to the trimer-hexamer equilibrium study, phycocyanin's hydrophobic interaction with different hydrocarbon chain lengths has been explored. Batch calorimeter experiments at 25°C demonstrated that the alcohols' hydrophobic binding capacity with phycocyanin was equivalent to their self-association capacity. The pairwise negative Gibbs energy of self-association becomes less favorable with decreasing alkyl chain length, and cyclohexanol exhibits the strongest tendency for self-interaction: cyclohexanol> butanol > propanol> ethanol> methanol> ethylene glycol. The pairwise interaction enthalpies of the alcohols are positive and decrease in the order butanol> propanol> ethanol = methanol. Because the measurements were made in such low quantities, it is hard to compare the findings directly.

Comparing polyols to alkyl chain-containing alcohols, the stabilizing impact of polyols on proteins has long been known. Polyol-induced stability may be explained using scanning microcalorimetry and density measurements, which helped to identify preferred interaction parameters.

Lysozyme's transition temperature was lowered by 17.8°C and 1.1°C, respectively, while its transition enthalpies were enhanced by 33.5 kJ/mol and 126 kJ/mol when methanol and ethylene glycol were 30 percent w/w.

Lysozyme's transition temperature and transition enthalpies increased when glycerol, erythritol, xylitol, and sorbitol were added to the solution at 10 to 50 w/w%. One reason is that all three thermodynamic transition parameters (AGo, Ahah, AS0) are increasing simultaneously. Sorbitol and glycerol concentration functions suggest the stabilizing effect is due to increased interaction enthalpy when the polyols are present [15].

Resonance Light Scattering (RLS)

Light scattering can be divided into three domains based on a dimensionless size parameter, a, which is defined as:

$$a = \frac{\pi DP}{\lambda} \tag{12}$$

Where πDp = circumference of a particle, and λ = wavelength of incident radiation. Based on the value of these domains are:

a « 1: Rayleigh scattering (small particle compared to the wavelength of light);

a ≈ 1: Mie scattering (particle about the same size as the wavelength of light); and,

a » 1: Geometric scattering (particle much larger than the wavelength of light).

Suspended particles exhibit random motion due to their interactions with suspending fluid molecules. The viscosity of the suspending fluid, temperature, electrical charge, electrical mobility, and particle size are all factors in Brownian motion according to the Stokes-Einstein theory, as shown in Equations (13) and (14):

$$D = \frac{K_B T}{6\pi \eta r} \tag{13}$$

$$D = \frac{\mu q K_B T}{q} \tag{14}$$

Where D = diffusion constant, q = electrical charge of a particle, μ_q = electrical mobility of the charged particle, kB = Boltzmann's constant, T = absolute temperature, η = viscosity, and r = radius of a spherical particle.

Equation (13) is the electrical mobility equation, and it shows how charged particles move through a certain medium. Lastly, the way particles move in a fluid of a certain temperature and viscosity can be used to figure out their size. The Mie scattering optical model is used in laser light scattering (LLS). DLS is a way to find out if there is optical motion. Coherent light is used to illuminate the suspended particles. Because the location or speed of the suspended particles changes over time, this causes the frequency of the light they scatter to change. Theoretically, a single spherical particle can make light and dark concentric bands. As you move away from them, they become less bright.

Classical and dynamic light-scattering methods are two of the most common. This model considers light scattering (absorption, refraction, and reflection) around the particle in the medium. The Mie Model using particle diameter and the refractive index function, we may get an approximation for the dimensionless size parameter. Optical frequency changes recorded over time are a function of random particle motion, which yields a full distribution [16].

Synchronous Fluorescence Spectra (SFS)

For the first time, Lloyd used pyrolysis and combustion of organic materials to produce benzo(k)fluoranthene, benzo(u)-pyrene, and perylene, a combination of C20 polynuclear hydrocarbons. This comparison of 23-nm interval synchronous scanning with optimal fixed excitation wavelengths yielded significantly different emission spectra.

Forensic research was the first area where the approach was used experimentally. Non-specialists, on the other hand, struggled to grasp and use the advantages of this simple method because of a lack of expertise and skill. As a consequence, Vo_{Dinh} provided essential theoretical and practical methodologies for analyzing spectral signatures from challenging samples and other analytically significant data. The intensity of synchronous fluorescence, Is, depends on the shape of the normal excitation and emission spectra, as well as the difference in wavelength, $\Delta\lambda$, between the excitation and emission wavelengths, $\Delta\lambda_{ex}$ and $\Delta\lambda_{em}$.

This intensity can be expressed as:

$$I_s = K_c dE_x(\lambda_{ex})E_m(\lambda_{ex} + \Delta\lambda) \tag{15}$$

or, alternatively

$$I_S = K_c dE_x(\lambda_{em} - \Delta\lambda)E_m(\lambda_{em}) \tag{16}$$

Where E is the excitation function at a given excitation wavelength ($\lambda_{ex} = \lambda_{em} - \Delta\lambda$), E is the normal emission intensity at the corresponding emission wavelength ($\lambda_{ex} = \lambda_{em} + \Delta\lambda$), c is the analyte concentration, d is the thickness of the sample, and K is a characteristic luminescence constant comprising the "instrumental geometry factor" and related parameters. Fluorescence spectroscopy may be summarized as follows by Vo-Dinh. Because of the synchronous scanning, the signal may be thought of as either an excitation or an emission spectrum, depending on whether the excitation or emission wavelengths are shorter or longer. Excitation and emission wavelengths, $\Delta\lambda$, are used to calculate the spectrum. A single component's maximum fluorescence intensity occurs when $\Delta\lambda$ the absorption and emission maxima of that component are equal.

a. Spectral band narrowing. This is a consequence of the multiplication of two functions that are rising or decreasing at the same rate.
b. Emission spectra have been reduced in complexity. Conventional fluorimetry can only raise the intensity of all emission bands at a given &, but synchronous fluorimetry allows for selective enhancement of the strongest peaks by utilizing an appropriate $\Delta\lambda$.
c. Reduced spectral coverage. Some features of the spectrum may not be of importance from an analytical standpoint; their presence only confuses the whole spectrum due to their overlap with other components in a sample's emission. The spectral bandwidth of the synchronous signal may be altered experimentally by adjusting $\Delta\lambda$ or the Stokes shift. Because of distortions generated by intermolecular interactions and static and dynamic quenching processes, synchrotron spectrometry is not without its drawbacks.

A recent paper examined the method's properties and potential uses. The peak wavelengths and intensities of synchronous fluorescence spectra may be predicted using a theory proposed by Lloyd and Evett. A decent degree of precision may be achieved if the peak maxima of the excitation and emission peaks are represented as Gaussian peaks. In the synchronous spectra of 33 compounds, the mean discrepancy between calculated and observed peaks was 1.95 nm. The synchronous excitation approach enhances the measurement over traditional techniques if proper A1 values are used for Rayleigh and Raman perturbation of the fluorescence of a molecule. Phenol in aqueous solutions has been measured experimentally, and this is the case.

The most challenging part of using the synchronous scan approach is determining the ideal $\Delta\lambda$ value. The use of several $\Delta\lambda$ values may be required in certain multi-component systems to ensure comprehensive identification. A three-dimensional graphic makes it simple to identify the optimal $\Delta\lambda$ values. Using three-dimensional spectrin to represent the excitation, emission, and intensity of a chemical's fluorescence, one axis may be used to depict the whole activity.

Excitation-emission matrix spectra, contour spectra, and total luminescence spectra are a few examples of these spectra. Fluorescence may be quantified using a computer-controlled apparatus that can automatically acquire the entire luminescence spectrum from emission and excitation wavelengths measured in only 16.7 seconds at a spatial resolution of one nanometer per point. Visualizes the intensity profile of the 3D plot at a 45-degree angle in three dimensions using a three-dimensional synchrophasor spectrum. Synchronous luminescence methods use mechanical scanning and simultaneous data capture, although diagonal scans of the spectrum matrix offer similar spectral information. It has been shown by Weiner that scanning three-dimensional fluorescence spectra are similar to scanning two-dimensional fluorescence spectra once data has been captured by the software.

The synchronous spectrum, which is more ambiguous than the total luminescence spectrum, may be readily captured by a simple fluorimeter that scans both the excitation and emission monochromators at the same time. For the same price as a regular fluorimeter, you can buy a video fluorimeter. Synchronous excitation's simplicity of implementation is one of its most enticing features.

When employing a commercial or laboratory-built luminescence detector, synchronized measurements are possible. Total luminescence information is lost when using the synchronous luminescence scanning approach. It is a straightforward and efficient way to collect data on several substances in a single experiment. White has boosted the sensitivity of a fluorescence spectrophotometer by thirtyfold by adding concave retro reflectors behind the sample in each beam and rotating the monochromator slit images by 90°. A rhodamine analytical calibration of 10^{-6} down to 10^{-12} g/ml was obtained using this device and the synchronous excitation approach [17].

Resonance Energy Transfer Efficiency

Understanding the function of macromolecular complexes requires knowledge of their structural and kinetic properties. Analytical biochemistry is interested in the phenomena of resonance enemy transfer as determined by steady-state and time-resolved fluorescence. The approach is well-suited to many biological systems since it can measure distances between 10 and 100 A. Perrin discovered the Phenomenon of resonance energy transfer in the early 1900s. However, it was not until the late 1940s that Forster offered a theory defining long-range molecule interactions through resonance energy transfer and developed an equation for the transfer rate. Stryer and his colleagues have proven the notion *via* experimentation. The transfer rate was shown to be influenced by distance and overlap integrals. In biology, a spectroscopic ruler based on resonance energy transfer can measure long distances.

Suppose the probe linker arm is flexible enough. In that case, the orientation factor does not significantly affect average distance measurements in many applications. Orientation ambiguity has been a key drawback in the use of this approach. Several options were discussed for reducing the problem. Now that we know how to solve the issue, we can proceed. Measurements of the apparent distance distribution may reduce the considerable degree of uncertainty. As a result, resonance energy transfer measurements may be used to estimate average distances.

There are various characteristics in the field of resonance energy transfer research. (a) Fluorescence detection is used to measure energy transfer, resulting in great sensitivity. Flow cytometry, electrophoresis, microscopy, *in vivo* detections, and liquid chromatography tests, to name a few, may all use energy transfer technologies in addition to the traditional Spectroscopic fluorometer for data collection. (b) Regardless of the complexity of the system's heterogeneity, structural information may be acquired at a low resolution. (c) Due to the nanosecond time scale of resonance energy transfer, many time-averaged processes in other approaches may be resolved (for example, the gradual conversion of conformers). Resonance energy transfer is the subject of many studies, but little is known about the method's actual use. Biochemistry (sample separation and purification), chemistry (chemical modification with appropriate fluorophores), physical measurements (steady-state or time-resolved), and data processing are all required for the method's use (usually nonlinear least squares). Resonance energy transfer technologies may now be effectively applied to biological challenges because of recent breakthroughs in molecular biology, purification procedures, instruments, and computer data processing. Forster distance estimation, sample preparation, energy transfer efficiency measurements by steady-state and time-resolved fluorescence, and data analysis are all included in this study [18].

UV-Vis Spectroscopy

The variations in the NPs' absorbance spectra caused by protein binding may be utilized to assess the strength of the binding. Dimension, aggregate formation, and local dielectric environments all impact how much the absorbance spectrum of this complex changes. During an investigation on SWCNT–BSA binding, scientists discovered that their complex's absorption spectrum was identical to the overlap between the SWCNT and BSA spectra. Carbon nanotubes, on the other hand, have a plasmon band that shifts and widens more often than other NPs.

It is possible to compute a constituent's concentration using Beer-Lambert law Equation (17) when the concentration of an additional substance is known:

$$A_\lambda = \varepsilon_{\lambda.a}.l.C_a + \varepsilon_{\lambda.b}.l.C_b \tag{17}$$

An SWCNT–BSA sample was compared to SWCNT–BSA samples in terms of spectra (continuous curves) and diamonds (diamonds). Calculated and observed spectra were found to be quite similar. As shown above, UV-vis spectroscopy can analyze NP–protein bindings. UV-vis is quicker, more adaptable, and less complex than other approaches.

The absorption spectra of various NPs, on the other hand, may reveal distinct properties. A different analytic procedure must be used if UV-vis alone does not provide clear data [19].

Conformational Changes of Proteins on the Surface of NPs

The surface chemistry of NPs (NPs) affects the structural alterations in the bound proteins. Size, shape, curvature, and surface area all influence the structural properties of bound proteins. Due to their surface properties, structural changes may be induced by binding to proteins. Corona proteins' structural alterations have biological significance. The disruption of physiological homeostasis and unwelcome immunological responses may be triggered by structural alterations that result in the loss of function.

Another factor that may lead to protein aggregates and amyloid fibers is alterations in the bulk structure of proteins. Peptides with intact Aβ40 peptides may be more likely to collide with one other when hydrogenated NPs are present. Using silica NPs, lysozyme from hen egg whites was destabilized, resulting in fibrous protein aggregation. To understand how NPs work, we must know how protein coronas in NP systems alter their shape [9].

Circular Dichroism (CD)

Spectroscopic methods such as FTIR, Raman, and CD are useful additions to protein secondary structure. Proteins' interactions with other molecules benefit greatly from this technique. CD spectroscopy uses absorption in the UV spectrum in near (visible at 250 nm) and far (180-250 nm) regions. Protein secondary and tertiary structures are revealed following binding. When studying the secondary structure, far-UV CD spectroscopy is often utilized [20].

Fourier Transform Infrared Spectroscopy (FT-IR) Measurements

There are a number of techniques for obtaining protein and other excipients' vibrational spectra using Fourier transform infrared (FTIR). It is possible to understand better how proteins interact with one other and excipients by doing FTIR research. A simple experiment may characterize bond lengths and strengths, bond angles and conformation, hydrogen bonds, electric fields, and conformational flexibility.

Characterization of protein structural content is the most prevalent use of FTIR in protein–excipient research. The protein secondary structure is defined by the Amide I vibrational bands. This band is made up of the carbonyl stretching vibration, as well as NH in-plane bend and out-of-phase CN stretching vibration. Bond coupling, hydrogen bonding, transition dipole coupling, and dipole-dipole interactions also impact the Amide I band.

An Amide I form *via* the bond is not affected by the coupling, but local vibrations may be affected. Because of this, if any of these interactions are disrupted, the spectrum might vary. Amide I frequency decreases by 25 cm^{-1} when hydrogen bonding to the carbonyl group is included, but it increases by 15 cm^{-1} when hydrogen bonding to the NH group is included. The intermolecular structure, intramolecular anti-parallel β-sheets, alpha-helices, and the 310-helix structure are all weaker than the protein's secondary structure in terms of hydrogen bonding strength. An interaction between two neighboring oscillating dipoles that are firmly linked when their oscillation frequencies are equal, known as transition dipole coupling (TDC), has a strong influence on the structure of the Amide I band.

In-plane bend NH and stretching CH create vibrations in the Amide II band, most often detected at 1550 cm^{-1}. Despite the fact that amino acid sidechains are not absorbed, Amide II is not strongly related to protein secondary structure. The

Amide II band is an excellent tool for studying H-bonding or isotopes. Additionally, the Amide III band is generated by one last NH bend and CH stretching vibration.

This spectral band, which ranges from 1200 to 1400 cm^{-1}, is made up of the side chains of amino acids and the secondary structure of proteins. Calibration sets of proteins with known structures and FTIR patterns are typically used to determine the secondary structure makeup of proteins using FTIR. As part of fitting band operations, the wide Amide I band is broken down into smaller sub-bands. By using techniques like Fourier self-deconvolution, fine structure enhancement, and second derivative analysis to restrict the bandwidth, it is possible to get more exact band placements. Gaussian or Lorentzian functions are used in order to fit each augmented band with a second structural unit. Once the structural unit has been established, the relative quantities of each secondary structure may be calculated. Calibration sets are another frequent method for estimating secondary structure composition from protein FTIR spectra.

Secondary structural correlations and IR spectra for a large number of proteins have been generated. The secondary structure content of unknown spectra may be deduced from the known secondary structure content of the baseline spectra. Prior to statistical analysis, the number of spectra may be reduced using principal component analysis and singular value decomposition. Secondary structure studies and the environment surrounding particular amino acid sidechains may both benefit greatly from protein IR spectroscopy. While most amino acids' protonation state and hydrogen bonding features are difficult to get experimentally, other amino acids are more accessible. Carboxyl groups' protons may be seen at a wavelength of 1710–1790 cm^{-1}, whereas the side chain of Cys can be seen at 2550–2600 cm^{-1}. In order to determine the protonation state, H-bonding properties, and relative hydrophobicity of the environment around these residues, a distinctive aromatic amino acid band may be employed, for example. To be safe, it is best to take care of and consider other options.

An additional experiment is needed to confirm the participation of individual residues since several of the side chain absorbance bands overlap (*e.g.,* isotope exchange).

It is possible to evaluate a wide range of excipient effects using FTIR spectroscopy. These include excipient-induced changes in protein structure and surface adsorption, excipient-protein interactions and coordination and chelation site changes, as well as excipient-induced hydration effects (*e.g.,* preferential water uptake) (*e.g.,* increased thermal unfolding temperature, Tm) [21].

Nuclear Magnetic Resonance Spectroscopy (NMR)

An NMR spectrometer employs the nucleus' magnetic properties to detect the chemical surrounds of a molecular structure without causing any damage or invasiveness. NMR may provide information on nanomaterials' structure, content, purity, molecular weight, dynamics, and diffusion characteristics in liquid and solid states in one-, two- and multi-dimensional investigations. The most recent breakthroughs in NMR spectroscopy also allow the investigation of nanomaterials in suspension and colloidal form.

Spectroscopy of nuclear magnetic resonance (NMR) is based on the nuclei's magnetic characteristics. Atomic nuclei in a magnetic field interact with radio frequency (MHz) radiation. These subatomic particles may be pictured whirling around their axes like a spinning top. Magnetic dipoles are created in rotational systems when the positively charged rotating nucleus functions as an effective bar magnet due to charge circulation. Nuclei with odd masses or atomic numbers are given the nuclear spin quantum number m. Nuclei having an I=1/2 spin number include 1H, 13C, 15N, 19F, and 29Si. An even-mass and even-number nucleus (12C, 16O) has a nucleus spin I=0 and is non-resonant in the NMR wavelength range. For nuclei with spins of 2I + 1, I assume the 2I + 1 orientation when a magnetic field is applied. Each of these directions corresponds to a different energy level denoted by a different magnetic quantum number (m = -I to I).

For instance, a nucleus with nuclear spin 1/2 may be oriented in one of two ways. In terms of the external magnetic field, one is perpendicular (m=1/2), while the other is parallel (m=+1/2). These orientations correlate to energy in the absence of an external magnetic field. They are only separated in energy when an external magnetic field is introduced. More stable states have a higher positive m value than unstable ones. To find out how much energy each spin state has, one may use the following equation:

$$E_i = m_i \frac{\gamma h B_0}{2\pi} \tag{18}$$

Planck's constant (h) and the magnetogyric ratio (γ) Depend on the external magnetic field; this constant varies with each isotope's spin state energies. The greater the applied magnetic field intensity, the greater the energy gap between nuclear spin states. A nucleus with an I=1/2 spin has a nuclear spin energy differential that may be calculated as follows:

$$\Delta E = \frac{\gamma h B_0}{2\pi} \tag{19}$$

An angle is present in this rotation of the spin nucleus with respect to the direction of the external magnetic field, despite the fact that it is not precisely parallel or anti-parallel. The periodic wobbling motion is known as precession. A nucleus' gyromagnetic ratio (γ) and the applied field's intensity (E) determine the "Larmor frequency," which is the precession frequency.

$$\omega = \gamma B_0 \tag{20}$$

Temperature, magnetic field strength, and the gyromagnetic ratio all influence the Boltzmann distribution of spin energy levels.

The energy difference between the two spin states is negligible. There are several nuclei in higher-energy spin states because of thermal collisions. Signal intensity is defined by the change in population between two energy levels when employing spectroscopy. In nuclear magnetic resonance, nucleus populations in various spin states have relatively little difference. NMR has a very small energy gap between spin states. There is a limit to the magnetic field's intensity in any case. As a result, a technique with lower sensitivity than NMR is selected. When a sample is placed in a magnetic field and exposed to radio waves, its nuclei absorb energy and change the orientation of nuclear spins. The precessional frequency must match the radiation in order to absorb energy.

Nuclear energy is liberated when the applied field ceases to act on the nuclei. Measurement of the molecule's energy output is used to construct the fingerprint spectrum. Spectroscopy using the Fourier transform and continuous-wave NMR is the two most often utilized NMR procedures. Radiofrequency (RF) pulses are used in Fourier transform NMR spectrometers for measurements. Samples in the liquid state are frequently analyzed using NMR spectroscopy. Recent years have seen a great deal of progress in NMR solid-state studies, including CP, the magic angle spinning, high-resolution magic angle spinning, and the magic angle spinning with cross-polarization. Its RF source and magnetic field are critical to an RF NMR spectrometer's performance. There are two magnet poles sandwiching a sample between them. Assume that the RF irradiation frequency and magnetic field intensity remain constant. If this is true, then the field strength at each nucleus varies somewhat as it approaches resonance. It is possible to measure the amount of RF energy absorbed by a very sensitive detector.

Also developed for use with other techniques like magnetic resonance imaging or scanning are methods such as heteronuclear single and multiple bond quantum correlations, total quantum coherence, exchanging quantum coherence, and diffusion-ordered quantum spectroscopy (DOSY).

Inorganic-organic composite NPs, block copolymer-based micellar NPs, dendritic NPs, nanocrystalline hydroxyapatite zinc oxide NPs, and carbon nanotubes may all be studied by NMR. Analysis of the CO bond's 13C chemical shift from 180.7 to 173.4ppm by NMR revealed significant interactions between ZnO and the chain's carboxyl groups. Quantum dots' chemical composition has been determined using CP-MAS-DNP NMR using 113Cd, 77Se, and 133Cs. NMR 1H-1H COSY, 13C-1H HSQC, and 1H DOSY may also identify the various kinds of organic compounds linked to NPs. An amphiphilic di-block copolymer's critical micelle concentration (CMC) may be determined using DOSY. The quantity of encapsulated small molecules inside NPs can be estimated using DOSY [22].

CONCLUSION

Characterization and analysis of proteins bound to the NP surface is the first step towards understanding the true nature of the NP-mediated biological effects. Research thus far highlights that size, shape, and surface characteristics of NPs affect protein adsorption and also have the capability to modify the structure of the adsorbed protein molecules. This can significantly affect the reactivity of the NP with cells and determine the route and efficiency of NP uptake. The adsorbed proteins may also promote translocation of the NP across cellular barriers and clearance or accumulation in vital organs. Interestingly, most studies conducted in this direction focus on *in vitro* test systems; therefore, extrapolation of this information in predicting the behavior of NPs *in vivo* remains a challenging task and needs further investigation. Systematic analysis of binding characteristics of novel NPs with proteins having different structures, shapes and functional properties can enhance our existing knowledge about NP-protein interactions. A thorough understanding of NP-protein interactions might lead to strategic manipulation of NP surfaces to adsorb specific functional proteins or small drug molecules intended for delivery *in vivo*.

Furthermore, this knowledge might also be useful in predicting nanotoxicity-related safety concerns. In summary, NP-PC dictates the overall biological reactivity of the otherwise inorganic NP surface. Understanding the dynamics of this complex interaction can thus provide useful insights into the cytotoxic, inflammatory potential and other key properties of these novel materials that can be explored for developing safer and value-added nanomaterials for future applications.

REFERENCES

[1]   I. Lynch, and K.A. Dawson, "Protein-nanoparticle interactions", *Nano Today,* vol. 3, no. 1-2, pp. 40-47, 2008.
        [http://dx.doi.org/10.1016/S1748-0132(08)70014-8]

[2]   M. Mahmoudi, I. Lynch, M.R. Ejtehadi, M.P. Monopoli, F.B. Bombelli, and S. Laurent, "Protein-

nanoparticle interactions: opportunities and challenges", *Chem. Rev.,* vol. 111, no. 9, pp. 5610-5637, 2011.
[http://dx.doi.org/10.1021/cr100440g] [PMID: 21688848]

[3] S.R. Saptarshi, A. Duschl, and A.L. Lopata, "Interaction of nanoparticles with proteins: relation to bio-reactivity of the nanoparticle", *J. Nanobiotechnology,* vol. 11, no. 1, p. 26, 2013.
[http://dx.doi.org/10.1186/1477-3155-11-26] [PMID: 23870291]

[4] F. Reactions, "Biophysical Principles", *Cell Biol. (Henderson NV),* pp. 53-62, 2017.
[http://dx.doi.org/10.1016/B978-0-323-34126-4.00004-9]

[5] S. Pylaeva, M. Brehm, and D. Sebastiani, "Salt bridge in aqueous solution: strong structural motifs but weak enthalpic effect", *Sci. Rep.,* vol. 8, no. 1, p. 13626, 2018.
[http://dx.doi.org/10.1038/s41598-018-31935-z] [PMID: 30206276]

[6] M. De, C.C. You, S. Srivastava, and V.M. Rotello, "Biomimetic interactions of proteins with functionalized nanoparticles: a thermodynamic study", *J. Am. Chem. Soc.,* vol. 129, no. 35, pp. 10747-10753, 2007.
[http://dx.doi.org/10.1021/ja071642q] [PMID: 17672456]

[7] P. Pino, B. Pelaz, Q. Zhang, P. Maffre, G.U. Nienhaus, and W.J. Parak, "Protein corona formation around nanoparticles – from the past to the future", *Mater. Horiz.,* vol. 1, no. 3, pp. 301-313, 2014.
[http://dx.doi.org/10.1039/C3MH00106G]

[8] Y. Kim, S.M. Ko, and J.M. Nam, "Protein-nanoparticle interaction-induced changes in protein structure and aggregation", *Chem. Asian J.,* vol. 11, no. 13, pp. 1869-1877, 2016.
[http://dx.doi.org/10.1002/asia.201600236] [PMID: 27062521]

[9] S.J. Park, "Protein–nanoparticle interaction: Corona formation and conformational changes in proteins on nanoparticles", *Int. J. Nanomedicine,* vol. 15, pp. 5783-5802, 2020.
[http://dx.doi.org/10.2147/IJN.S254808] [PMID: 32821101]

[10] J. Haginaka, "8.9 chromatographic separations and analysis: protein and glycoprotein stationary phases", In: *Nuclear Chemistry* vol. 8. , 2012.
[http://dx.doi.org/10.1016/B978-0-08-095167-6.00822-3]

[11] F. Babick, "Chapter 3.2.1 - Dynamic light scattering (DLS)", In: *Characterization of Nanoparticles* Elsevier Inc., 2019, pp. 137-172.
[http://dx.doi.org/10.1016/B978-0-12-814182-3.00010-9]

[12] S. Bhattacharjee, "DLS and zeta potential – What they are and what they are not?", *J. Control. Release,* vol. 235, pp. 337-351, 2016.
[http://dx.doi.org/10.1016/j.jconrel.2016.06.017] [PMID: 27297779]

[13] S. Singh, and B.S. Tiwari, "Chapter 4 - Biosynthesis of High-Value Amino Acids by Synthetic Biology", In: *Current Developments in Biotechnology and Bioengineering.* Elsevier B.V., 2018, pp. 257-294.
[http://dx.doi.org/10.1016/B978-0-444-64085-7.00011-3]

[14] K.K. Turoverov, and I.M. Kuznetsova, "Intrinsic UV-fluorescence of proteins AS a tool for investigation of their dynamics", *Tsitologiya,* vol. 40, no. 8–9, pp. 745-746, 1998.
[PMID: 9821244]

[15] H.J. Hinz, "Thermodynamics of protein-ligand interactions: calorimetric approaches", *Annu. Rev. Biophys. Bioeng.,* vol. 12, no. 1, pp. 285-317, 1983.
[http://dx.doi.org/10.1146/annurev.bb.12.060183.001441] [PMID: 6347040]

[16] S.K. Brar, and M. Verma, "Measurement of nanoparticles by light-scattering techniques", *Trends Analyt. Chem.,* vol. 30, no. 1, pp. 4-17, 2011.
[http://dx.doi.org/10.1016/j.trac.2010.08.008]

[17] S. Rubio, A. Gomez-Hens, and M. Valcarcel, "Analytical applications of synchronous fluorescence spectroscopy", *Talanta,* vol. 33, no. 8, pp. 633-640, 1986.

[http://dx.doi.org/10.1016/0039-9140(86)80149-7] [PMID: 18964158]

[18] P.G. Wu, and L. Brand, "Resonance energy transfer: methods and applications", *Anal. Biochem.,* vol. 218, no. 1, pp. 1-13, 1994.
[http://dx.doi.org/10.1006/abio.1994.1134] [PMID: 8053542]

[19] L. Li, Q. Mu, B. Zhang, and B. Yan, "Analytical strategies for detecting nanoparticle–protein interactions", *Analyst (Lond.),* vol. 135, no. 7, pp. 1519-1530, 2010.
[http://dx.doi.org/10.1039/c0an00075b] [PMID: 20502814]

[20] J.J. Sutkovic, "A review on Nanoparticle and Protein interaction in biomedical applications", *Periodicals of Engineering and Natural Sciences (PEN),* vol. 4, no. 2, 2016.
[http://dx.doi.org/10.21533/pen.v4i2.62]

[21] T.J. Kamerzell, R. Esfandiary, S.B. Joshi, C.R. Middaugh, and D.B. Volkin, "Protein–excipient interactions: Mechanisms and biophysical characterization applied to protein formulation development", *Adv. Drug Deliv. Rev.,* vol. 63, no. 13, pp. 1118-1159, 2011.
[http://dx.doi.org/10.1016/j.addr.2011.07.006] [PMID: 21855584]

[22] M. Kaliva, and M. Vamvakaki, "Pavement types, wheel loads, and design factors", In: *Principles of Pavement Design* vol. 1242. Second Edition. Elsevier Inc., 2010, pp. 1-23.
[http://dx.doi.org/10.1002/9780470172919.ch4]

CHAPTER 5

Toxicity of Nanomaterials-Physicochemical Effects

Abstract: Nanoparticles (NPs) have the potential to produce deleterious effects on organ, tissue, cellular, subcellular, and protein levels due to their peculiar physicochemical features. Metal NPs are gaining prominence and are being used in a variety of medicinal, consumer, industrial, and military applications. Furthermore, as particle size falls, some metal-based NPs become increasingly poisonous, despite the fact that the same substance is rather innocuous in its bulk form. NPs can also interact with proteins and enzymes within human cells, causing reactive oxygen species to be produced, an inflammatory response to be initiated, and mitochondrial disruption and destruction, ending in apoptosis or necrosis. As a result, deciding whether the advantages of NPs outweigh the hazards presents various challenges.

Keywords: Cellular toxicity, Cytotoxicity, Nanotoxicity, Nanoparticles toxicity.

INTRODUCTION

Muller's (1927) work on "artificial transmutation of the gene" in the fruit fly, Drosophila melanogaster, prompted interest in exploring harmful effects on inherited genetic information in cells. The earliest scientific study on chemically induced mutation used Muller's fruit fly model to describe the mutations caused by exposure to sulfur mustard. In 1966, geneticists at a meeting sponsored by the National Institutes of Health in the United States proposed that food additives, medications, and substances with extensive human exposure be systematically examined for mutagenicity. NPs are frequently employed in electronics, agriculture, textile manufacture, medicine, and other sectors and sciences. The key obstacle restricting their application in disease therapy and detection is NP toxicity in live organisms. At the moment, researchers commonly encounter the challenge of reconciling the favorable therapeutic impact of NPs with the toxicity-related adverse effects. Choosing an appropriate experimental paradigm for assessing toxicity *in vitro* and *in vivo* is crucial in this respect. Whereas *in vivo* tests enable you to estimate the NP toxicity for particular organs or the body as a whole, *in vitro* models make it simpler to assess the NP's harmful effects on individual cell components and tissues. The concentration, length of their contact with living things, stability in biological fluids, and capacity to build up in tissues and organs are other factors that affect NPs' potential toxicity. Understanding the

Seyed Morteza Naghib and Hamid Reza Garshasbi

connections between all the variables and the processes driving NP toxicity is essential for the creation of safe, biocompatible NPs for the detection and treatment of human illnesses.

The production, use, disposal, and waste treatment of products containing NPs are the primary causes of their release into the environment. Typically, the epidermis protects against external chemicals, whereas the lungs and digestive system are vulnerable organs. NPs are around the size of viruses. For instance, the diameter of the human immunodeficiency virus (HIV) particles is 100 nm [1].

The circulation and other organs, including the heart, liver, and blood cells, are simple entry points for inhaled NPs. It is crucial to understand that NPs' toxicity is influenced by their source. Some seem to provide health advantages, while others seem to be nontoxic [2]. The transfer of active chemical species through organismal barriers such as the skin, lungs, body tissues, and organs is made easier by the small size of NPs. Therefore, NPs can result in asthma, cancer, irreversible oxidative stress, organelle damage, and other conditions depending on their makeup. The formation of reactive oxygen species, protein denaturation, disruption of mitochondrial function, and alteration of phagocytic activities are only a few of the typical acute toxic consequences brought on by exposure to NPs and nanostructured materials. Common chronic harmful consequences of NPs include uptake by the reticuloendothelial system, nucleus, neuronal tissue, and the formation of neoantigens that could lead to organ growth and malfunction. The general characteristics of NPs that are used to categorize them include dimensionality, composition, morphology, aggregation, and homogeneity.

Like free NPs, nanostructured thin films and fixed nanoscale circuits found in computer microprocessors all have key distinctions that make it simpler to categorize them for specific applications. Free NP movement is unrestricted, which makes it simpler for them to spread across the environment and pose possible health problems when exposed to people.

On the other hand, handling fixed NPs properly poses no health dangers because the nanostructured components are affixed to a substantial object. An excellent example of a material whose initial states are safe is asbestos. The subsequent mining of asbestos results in the creation of nanoscale fibrous particles, which are then turned into an airborne aerosol, which is carcinogenic and poses serious health risks when ingested [1].

It is also important to remember that, in addition to size and age, the chemical makeup and form of the particle are the key determinants of nanoparticle toxicity. Many NPs in this situation are harmless, while others have diminished toxicity or could even have long-term negative health impacts [1].

Due to their propensity to penetrate cells and move within them, foreign NPs cause organelle injury or oxidative stress that permanently damages cells. Other than penetration, NPs bind to cellular components and kill cells by electrostatic charges, van der Waals forces, interfacial tension effects, and steric contact. Reactive oxygen species are produced by various NPs, which can then damage cells by altering lipids, proteins, DNA, signaling processes, and gene transcription [1]. The chemistry, shape, size, and placement of the NPs all affect how the oxidative products are disposed of. The cytoplasm, cytoplasmic components, and nucleus are only a few examples of diverse biological places where NPs might move or spread. Due to their cellular localization impact, NPs can damage cell organelles or DNA and result in cell death. According to toxicological data, the toxicity of NMs depends on various factors:

- Effect of exposure duration and dose. The molar concentration of NPs in the nearby media multiplied by the exposure period directly determines the number of NMs that enter the cells [1].
- Effect of accumulation and concentration. The toxicity of NPs at various concentrations has been the subject of numerous conflicting reports. Aggregation is encouraged by an increase in NP concentration. As most NP aggregates are only a few micrometers, their toxicity may be reduced because a sizable portion of them may not enter cells.
- The impact of particle size. The toxicity of NPs is size dependent. Ag NPs with a diameter of less than 10 nm have a stronger capacity to infiltrate and disturb the cellular systems of multiple species than Ag+ ions and Ag NPs with bigger diameters.
- Effect of particle shape. NPs have shape-dependent toxicity, meaning toxicity levels vary depending on the aspect ratio. For instance, asbestos fibers as short as 5 to 10 microns can produce mesothelioma, asbestos fibers as long as 10 microns can cause lung cancer, and asbestos fibers as long as 2 microns can cause asbestosis [3].
- Effect of surface area. The toxicological impact of NPs typically grows as surface area and particle size decrease. Also, it should be emphasized that human cells respond differently to nano and microparticles at the same mass dose.
- The impact of crystal structure. Depending on the crystal structure, NPs can exhibit a variety of cellular uptake, oxidative reactions, and subcellular localization. The toxicity of two crystalline polymorphs of TiO_2 varies, for example. In the dark, rutile NPs (200 nm) induce DNA damage by oxidation, but anatase NPs (200 nm) do not cause DNA damage [4].
- Effect of surface functionalization. The surface characteristics of NPs have dramatically impacted translocation and ensuing oxidation processes [5, 6].

- Effect of pre-exposure. Pre-exposure to lower NP concentrations or a shorter exposure duration can promote cellular phagocytic activity [1]. Because of this pre-exposure, the body becomes somewhat more tolerant to NPs [7].

MECHANISMS OF NANOPARTICLE TOXICITY

It is crucial to comprehend the main, but not the only, hazardous strategy by which NPs are dangerous before thinking about engineering them to lessen their impacts. The direct interaction of NPs with cell membranes, the dissolution and release of toxic ions, and oxidative stress are the three primary mechanisms by which they cause harm to live beings. After describing the pathways of toxicity, we offer mitigation measures for each mechanism to lessen toxicity and promote the long-term usage of nanomaterials [8].

Nanoparticle Binding to Cell Exterior

NPs can be hazardous when they come into touch with the surface of cells. In multicellular organisms, this interaction may lead to the uptake of NPs by a subset of cells; the effects of this have previously been covered. In addition to physically interfering with the bacterial surface by removing or destroying the lipid membrane, triggering internal signaling pathways that harm the cell, dissolving to release cell-permeable toxic ions straight at the bacterial surface, and opening up other toxic pathways, NPs are primarily found on the surface of bacteria [8]. The electrostatic attraction governs NPs interactions with cells and other living things. Positively charged NPs are anticipated to come into contact with bacteria more often than negatively charged NPs since bacteria typically have a negative surface charge. Gram-positive or Gram-negative bacteria were treated with gold nanoparticles (AuNPs) functionalized with either positive or negative charges of mercaptopropyl amine or poly (allylamine hydrochloride), PAH by Feng *et al.* [9].

Using biological TEM, they proved that the AuNPs were attached to the bacterial surfaces. It can be seen in certain photos that the AuNPs had successfully removed and attached entire lipid bilayers from the cell wall. They used flow cytometry to estimate the number of bacteria bound to gold in the area. The measured toxicity of each AuNP suggested that greater toxicity for both species resulted from enhanced nanoparticle binding. According to the research by Jacobson *et al.* [10], binding to the negatively charged surface molecule lipopolysaccharides dominates NP interaction with Gram-negative bacteria.

Once attached to the surface of bacteria, NPs can disrupt and harm cell membranes, potentially leading to cell death. AFM and quartz crystal microbalance with dissipation monitoring were utilized in investigations with poly-(diallyl dimethylammonium chloride)-coated CdSe quantum dots (QDs) to show how these QDs link to and embed in a supported lipid bilayer. They cause membrane damage by compressing the bilayer's liquid-ordered sections, which are required for membrane trafficking and signal transduction in both prokaryotic and eukaryotic cells. The SI also discusses other research works that investigated the toxicity of QDs with respect to model membranes and bacterial surfaces [11, 12].

Dissolution to Toxic Ions

It is unimportant whether the harmful substances in NPs disintegrate before or after sticking to an organism or the environment; this is a key process. Ions that have been released can be hazardous in a number of ways, depending on the nature of the ion. Ions can bind to proteins and enzymes, impairing their functioning and interfering with biological activities.

Additionally, direct interactions with metal ions can disrupt an organism's phospholipid membrane or genetic material. Ultimately, metal ions in organisms can produce oxidative stress. Much study has been conducted on the relevance of dissolving as a nanoparticle toxicity mechanism for many types of NPs. One such instance is discussed in the SI. Dissolution has been identified as a significant contributor to the toxicity of silver NPs. The bacteria *Shewanella oneidensis* MR-1 was found to be very hazardous to complex oxides such as lithium nickel manganese cobalt oxide (NMC) primarily by dissolving. Optical density measurements and respirometry were utilized by Hang *et al.* [13] to track bacterial growth and respiration, respectively. These investigations showed that NMC decomposed to release lithium, nickel, manganese, and cobalt ions but that the toxicity of the NMC nanomaterial was recapitulated following dosing with equivalent amounts of nickel and cobalt-released ions. To demonstrate the role of solubility in NMC toxicity, *equistoichiometric* NMC in three distinct morphologies was generated and tested against *S. oneidensis* [13]. These varied morphologies were chosen because they provide distinct crystal faces, even though it was anticipated that dissolution would vary based on the various crystal faces that gave differing degrees of transition metal coordination. This investigation revealed that transition metal coordination in different crystal faces might be crucial. Nanosheets, nano blocks, and a commercial, microscale NMC were the most dangerous nanomaterials when the dosage was based on mass. The NMC of all three morphologies was dosed by surface area as opposed to mass since dissolution is proportional to the material's exposed surface area. All morphologies displayed the same toxicity trend with surface area-based dosages, showing that any variations in dissolution depending on the presenting crystal face were not substantial enough to alter the reported toxicity to *S. oneidensis*.

The SI provides a thorough description of additional research examining the effects of dangerous NPs on dissolving [14, 15].

Mechanisms of Toxicity

The physicochemical reactivity of NPs causes free radicals or ROS, such as superoxide radical anions and hydroxyl radicals, to generate either directly or indirectly through the activation of oxidative enzymatic pathways, which results in oxidative stress. There are several sources of oxidative stress in general: 1. Oxidant-generating properties of particles, as well as their ability to stimulate ROS production as part of the cellular response to NPs. During the production of non-metal NPs, transition metal-based NPs or transition metal contaminants are used as catalysts. 2. Relatively stable free radical intermediates are found on reactive particle surfaces. 3. Redox-active groups formed as a result of nanoparticle functionalization [16].

Oxidative Stress

If not ubiquitous, oxidative stress is one of the most prevalent toxins that can cause tissue damage and harm the health of an organism.

Reactive oxygen species (ROS) generation and the oxidative stress it causes have both been extensively studied as a nanotoxicity mechanism. Lipid peroxidation is a frequently observed side effect of ROS production. Moreover, ROS have the ability to target specific important enzymes, including mononuclear iron proteins. In addition, ROS have the ability to oxidize DNA bases and deoxyribose, resulting in mutations and harm to the organism. This is clear from research showing that in aerobic settings, mutations were more common in mutants missing the enzyme machinery to remove ROS from cells [8].

ROS are formed when mitochondria and chloroplasts emit high-energy, uncoupled electrons during respiration. Electron leakage and oxidative damage may result from toxic chemicals' uncoupling electron transport in cells.

Oxidative metabolism of xenobiotics and inflammatory processes that produce nitric oxide and hydrogen peroxide can also produce reactive oxygen species. For a more in-depth look at the function of oxidative stress in pathogenesis, we recommend you check current reviews on environmental oxidative stress. Oxidative stress occurs when cells develop excessive amounts of oxygen radicals, which may surpass the usual antioxidant capability of the cell. Although antioxidants, such as tocopherols, ascorbic acid, glutathione, and oxygen radical scavenger enzymes (catalase, peroxidase, and superoxide dismutase, SOD), can control reactive species concentrations, damaged (mutated) cells may still suffer from oxidative damage that leads to cell death, genotoxicity, and even cancer. Oxidative stress can be caused by the following factors: (1) the presence of xenobiotics, (2) immune system activity in reaction to invading microbes (inflammation), and (3) radiation, making oxidative stress a common denominator of toxicity or stress [17].

In connection to their electrical structures, several nanomaterials' toxicity and ROS production are being studied. Li *et al.* [18] used seven metal oxide NPs with band edges near the redox potential of reactive redox couples to show that the toxicity of a substance to Escherichia coli was inversely linked to the quantity of abiotic ROS generated. Seven toxicants were found in research that examined the E. coli toxicity of 24 metal oxide NPs. Only the hazardous NPs caused E. coli to create more intracellular ROS after being exposed to them, and five of the poisonous NPs also produced ROS on their own when bacteria were not present.

Using nanostructure activity connection analysis, it was shown that the toxicity of metal oxide NPs is connected with both the material's conduction band energy and its hydration enthalpy, which evaluates how quickly the material dissolves. The material was more flammable and more likely to dissolve if its conduction band energy overlapped that of biomolecules.

The results supported the hypothesized toxicity of certain nanomaterials and may help in the development of NPs that lessen their harmful effects. In the SI, the toxicity of metal oxide NPs is discussed in further detail. In addition to creating ROS on their own, nanomaterials can activate organisms' signaling pathways that cause oxidative stress.

In a study by Dominguez *et al.* [19], the effects of contact with nanodiamonds of two sizes, either positively or negatively charged, on oxidative stress in Daphnia magna guts were evaluated. This proved that, in the guts of D. magna, nanodiamonds with a diameter of 5 nm produced significantly fewer ROS than those with a diameter of 15 nm. The gut cells responded to the second encounter by expressing more oxidative stress genes, demonstrating that they were battling ROS.

Cytotoxicity

Cytotoxicity assays come in two different varieties: *in vivo* and *in vitro* research. *In vitro* toxicity tests are speedier, more useful, and less expensive and do not have any moral conundrums that come with *in vivo* toxicity tests' length of time, cost, and ethical conundrums. *In vitro* assays are the favored method for assessing the toxicity of the majority of nanomaterials because of these advantages [20]. *In vitro* methods are used to assess both the health of the cell membrane and the metabolic activity of living cells. Examining the integrity of the cell membrane is one of the most often used techniques for figuring out whether or not cells are viable. It is based on the evaluation of LDH activity in the extracellular medium and the leakage of substances like lactate dehydrogenase (LDH) that typically reside in cells. Alternately, the ability of dyes like trypan blue and neutral red to enter injured cells and stain intracellular components can be used to measure the integrity of the membrane. Living cells cannot be accessed by these colors.

Viable cells' metabolic activity can be measured using colorimetric assays like the MTT and MTS tests. Cell viability experiments commonly use luciferase, a protein that catalyzes the creation of light from adenosine triphosphate (ATP), to measure the number of surviving cells by watching the uptake and accumulation of neutral red dye and trypan blue after exposure to the toxin. The most popular *in vitro* methods for assessing the cytotoxicity of NPs are the LDH, MTT, and MTS assay [21 – 24].

The cytotoxicity of NPs coated on the surface must be tested before being used *in vivo* (*i.e.,* nontoxic). Numerous methods for measuring cytotoxicity have been developed so far. Assays such as ATP, MTT, lactate dehydrogenase (LDH), and 3-(4,5-dimethylthiazol-2-yl)-2,5-diphenyltetrazolium bromide are among the most often used ones. It is standard practice to evaluate aqueous sample solutions up to a concentration of 1 mM Ln.

Cell viability is dependent on maintaining a high enough concentration of it. For cell survival, ATP is critical during the ATP test. In addition, the number of cells in the body rapidly reduces when a cell is killed. To know how effectively NPs operate, you may use control cells (*i.e.,* cells not treated with NPs). The cells are injected with luciferase and D-luciferin to detect ATP levels. This process produces light as a byproduct:

$$ATP + D - LUCIFERIN + O_2$$

$$\xrightarrow{luciferase\ Mg^{2+}} oxyluciferin + AMP + PP_1 + CO_2 + light$$

Viability Detection Luminescent CellTiter-Glo aqueous sample solutions, including NPs, may be tested for cell toxicity using an assay (Promega, Wisconsin, USA).

By measuring the amount of light emitted, one may know how much ATP a cell has. NCTC1469 and DU145 prostate cancer cells contain gadolinium oxide NPs coated with D-glucuronic acid. D-glucuronic acid-coated gadolinium oxide NPs may be used in blood containing up to 200 micrograms of gadolinium oxide per milliliter without causing adverse effects. *In vitro* cell viability of more than 90% at a concentration of 1 mM or less is required for a sample to be considered biocompatible.

An important function of mitochondria in a living cell can be examined indirectly using the MTT technique. The cells are dyed yellow with an MTT stain. Succinate dehydrogenase in the cell's mitochondria becomes insoluble, dark purple formazan. Because this reduction process happens only in living cells, it measures cell viability. Isopropanol organic solvent, measured spectrophotometrically at 490 nm, extracts formazan from cells. If a cell's plasma membrane is breached, LDH, a soluble cytosolic enzyme, is released. There is a correlation between how many cells have died and how much LDH is produced.

Indirectly, LDH concentration is assessed. 4-iodophenyl) 3-(4nitrophenyl) 5-phenyl-2H-tetrazolium chloride are the four compounds that make up the test solution (tetrazolium salt INT). The results of the study will be examined. Lactate acts as a catalyst for converting NAD+ into NADH by LDH. NADH changes the salt INT into red formazan, which may be detected spectrophotometrically at a wavelength of 490 nm. Only live cells take up the neutral red in the neutral red method. There is a direct correlation between the number of live cells and the dye incorporation into the lysosomes. Before the cells may be employed again, the dye in the medium must be removed with many washes and an acidified ethanol solution. Spectrophotometric analysis is then utilized to determine the dye concentration in the sample [25].

Genotoxicity

Genotoxicity data have traditionally been used by national and international regulatory organizations as part of a weight-of-evidence (WOE) study to establish whether a medicine is likely to cause cancer. A few particular tests may determine a chemical's ability to cause cancer. If these tests are properly recorded, it is possible to conclude that the chemical has the potential to cause cancer. During embryogenesis and the development of the fetus, mutations of germ cells or genotoxicity in somatic cells may occur. Based on the data on mutagenicity, scientists can assess the risk associated with these occurrences.

Mitigating variables, such as toxicokinetic (*e.g.,* phenol and hydroquinone) or overwhelming toxicity, may reduce the risk of a mutagenic MOA in a chemical's carcinogenicity or other undesirable consequences (*e.g.,* dichlorvos). An endpoint for some regulatory agencies, such as those in the United States, Canada, the United Kingdom, and the European Union, is a heritable mutation. Increased heritability of illness is possible when a person's germ cells change over time. Cancer, sickle cell anemia, and neurological problems may be caused by mutations in the germline or somatic cells. The Human Gene Mutation Database compiles genetic mutations linked to human illnesses. When testing for mutagenicity, international guidelines should be followed when they exist. Mutagenicity was selected as one of the six key toxicity testing categories by the Oecd since it is one of the most prevalent causes of cancer. There are three main ways that genetic damage associated with human illness is evaluated when evaluating the mutagenicity of chemical substances:

1) Gene mutation; 2) Clastogenicity; and 3) Aneuploidy

Data about one molecule is likely to be based on recognized methodologies. Enzymes that include proteins and one or more low molecular weight chemicals, as well as pesticide metabolites that are safe for people and animals, are examples of this [26].

PROPERTIES THAT AFFECT TOXICITY COMPOSITION (METAL-BASE AND CARBON-BASED)

Metal-based NPs toxicity

Aluminum NPs

Soldiers and other military people are more likely to be exposed to Al NPs because of the variety of military applications for which they may be used, including coatings, propellants, and fuels.

Wagner and colleagues researched the effects of aluminum oxide and aluminum NPs on cell survival and cell phagocytosis. At 30 and 40 nanometers, aluminum oxide (Al_2O_3NP) was applied to rat alveolar macrophages (NR8383) for the first *in vitro* cellular effects of aluminum metal NPs with a 2–3 nanometer oxide cover (Al NP at 50, 80, and 120 nm). TEM, DLS, laser Doppler velocimetry, and/or CytoViva150 Ultra Resolution Imaging (URI) were used to analyze the nanomaterials in the lab and situ. According to DLS and URI results, the particles in cell exposure media were substantially larger than in TEM. According to cell viability tests, 24 hours of exposure to 100 g/mL of Al_2O_3 NP did not influence macrophage viability.

Although phagocytosis was unaffected by Al_2O_3 NP, other effects were as follows. It was found that 24 hours of exposure to Al NPs of various sizes (50, 80, and 120 nm) at the same concentration (25 g/mL) dramatically reduced cell phagocytosis but affected cellular viability. However, viability was drastically decreased after 24 hours of exposure to Al NP at dosages ranging from 100 to 250 g/mL. As a result of these experiments, Al NP is more hazardous to macrophages and reduces their phagocytotic function by a wider margin than Al_2O_3 NP after only 24 hours of exposure (Fig. **1**).

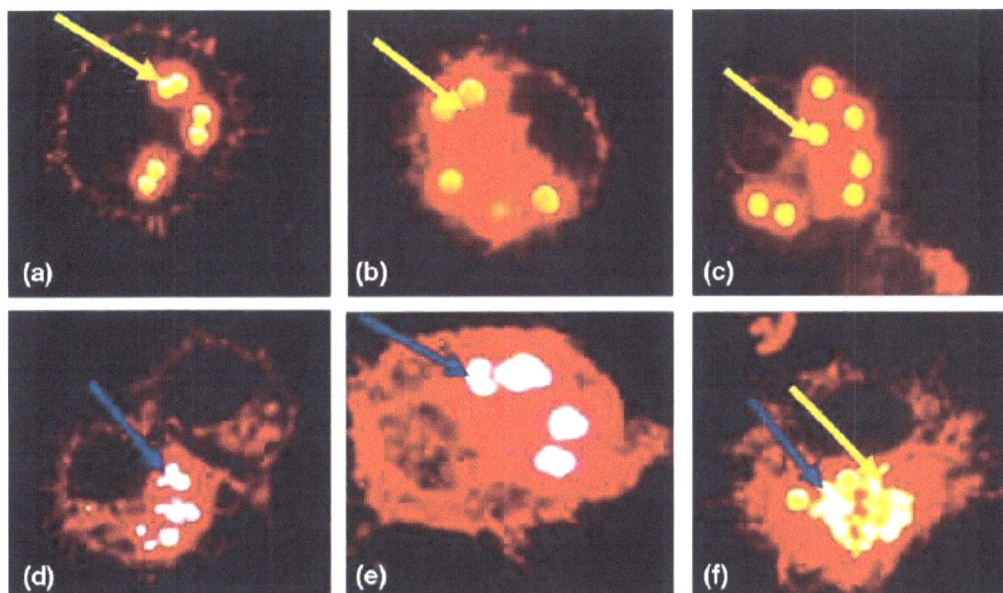

Fig. (1). Alveolar macrophage phagocytization of Al_2O_3 NPs (NPs) and AlNPs seen under the microscope (AM). Representative phagocytosis pictures (a–f) were captured using the Olympus IX71 inverted fluorescence microscope linked to an advanced high-illuminating system. Al_2O_3NPs and Al NPs were applied to cells for 24 hours at 5 or 25 g/mL concentrations. After exposure, cells were given 2 m-long fluorescent latex beads. A 10:1 dosage (10 beads per cell) was administered for six hours, and the beads emerged in bright globular portions of the cells. A phagocytosis index (PI) was calculated by counting macrophages and the beads they phagocytosed. The macrophage acceptance rate divided by the average number of beads accepted per positive macrophage is the formula used to calculate it.

Gold NPs

Apoptosis and LDH leakage at doses ranging from 1 to 100 g/mL after 48 hours of exposure to Al NPs were significantly increased in the presence of toxic effects on mammalian germline stem cells. Using four distinct cell lines, Pan *et al.* from the University of Texas at Austin studied the toxicity of gold NPs (0.8–15 nm) about their size. They discovered that the most lethal nanoparticle had the smallest molecular weight (1.4 nm). On the other hand, larger particles showed no harm, even at concentrations as high as 6.3 mg/mL. In the tests, gold nanoparticles (Au NPs) were used so that the effects of surface charge could be investigated on E. coli bacteria red blood cells and Cos-1 cell lines. Smaller (10–50 nm) than larger (100–200 nm) Au NPs were more hazardous in mice during an *in vivo* toxicity test using spherical colloidal suspensions. However, there was no specific information provided on the chemistry of the surface. Toxic Au NPs (2.8, 5.55, and 38nm in diameter) were created by grinding bulk gold, and the smaller Au

NPs increased the production of proinflammatory genes such as interleukin-1 (IL-1), interleukin-6 (IL-6), and tumor necrosis factor (TNF-α). According to Wang *et al.*, CTAB-coated gold nanorods were more toxic to human HaCaT keratinocytes than spherical gold NPs. The shape of the chemical affects its toxicity (30 nm).

Studies have demonstrated that positive-charged AuNPs are more harmful than negative-charged ones of the same size. This is in line with findings from past studies. Citrate reduction of gold NPs had little effect on lung fibroblasts, which measured an average of 20 nm in diameter. DNA had reached the point where it could no longer be repaired since it had been oxidized. According to a current study conducted on the topic, there was a reduction in the expression of genes that are related to DNA damage and the cell cycle.

In human dermal fibroblasts, 14-nm-diameter gold nanoparticles (Au NPs) alter actin and the extracellular matrix. The proliferation, adhesion, and mobility of cells are all hampered when faulty proteins are present. Au NPs have been proven to pass through the small intestine and reach circulation in several *in vivo* investigations. These viruses may reach various organs, such as the heart, lungs, hepatobiliary spleen, and digestive tract. The chorionic pore canals disperse single gold NPs throughout the chorionic space and into the embryo's core mass. Embryos were given 0.025–1.2 nm Au NPs for 120 hours. 74% of the treated embryos matured into normal zebrafish; 24 percent died, and two percent showed abnormalities. Despite the negative findings, multiple additional studies have shown that Au NPs do not damage cells or negatively affect the body. Surface-modified spherical AuNPs (4, 12, and 18 nm) were safe for human leukemia cells.

The proinflammatory cytokines TNF- and IL-1 were not secreted by macrophages after 72 hours of exposure to round, 3.5 nm gold NPs capped with lysine, as reported by Shukla *et al.* at concentrations up to 100 M. It is possible, based on the results of these two investigations, that the synthesis circumstances and the surface chemistry of gold NPs influence the biological response.

Silver NPs

Ag NPs, as antimicrobials, have been widely used in consumer and medical products, as previously mentioned. Signal augmentation and optical sensors, biomarkers, and *in vivo* imaging agents may all benefit from the plasmon resonant optical scattering features of Ag NPs. Ag NPs have the potential to produce (ROS) and oxidative stress, both of which may be detrimental to brain tissue and cell imaging since they may have a role in the development of neurodegenerative diseases such as Alzheimer's and Parkinson's.

A 24-hour exposure to Ag NPs resulted in oxidative stress in neuroendocrine cells, liver cells, lung cells, and germline stem cells, which may lead to cell death. Alveolar macrophages were studied by Carlson and colleagues, who discovered that Ag NPs were active throughout the body. The initial line of defense against foreign material in the lungs was alveolar macrophages. One of the probable explanations for their research was oxidative stress; *in vitro* treatment led to unusually big and adherent cells and considerable NP uptake at high dosages after 24 hours. Experiments in toxicology, which included measuring the viability of mitochondria and cell membranes and ROS, revealed that the toxicant concentration had a dose-dependent effect on cell viability. Oxidative stress may be the cause of the toxicity in cells exposed to 50 mM 15-nm Ag NPs since ROS levels rose more than 10-fold. There was also a check to see whether the inflammatory mediators TNF-, macrophage inhibitory protein (MIP-2), and interleukin-6 (IL-6) released into the culture medium were increased. IL-1β, MIP-2, and TNF-R (TNF-α) were all detected 24 hours after exposure to Ag NPs of 15 nm size. IL-6 levels, on the other hand, did not alter after exposure to Ag NPs.

The viability of utilizing Ag NPs as biomarkers was tested using Neuro-2A cells. Light microscopy indicated considerable optical labeling after 24 hours of incubation with strong illumination of hydrocarbon and polysaccharide Ag NPs, according to Schrand and colleagues. This was because both forms of Ag NPs excited plasmon resonance. In the same way, both kinds of Ag NPs that were affixed to the Neuro-2A cells' surface were ingested and transported into the intracellular vacuoles. ROS generation, loss of mitochondrial membrane integrity, and disruption of the actin cytoskeleton were seen after incubation with Ag NPs at concentrations of 25 g/mL or higher. Ag NPs made of hydrocarbons were shown to have the highest impact.

Sprague–Dawley rats received three doses of Ag NPs (60 nm), and their plasma alkaline and blood cholesterol levels changed significantly after 28 days. These three doses were 30 mg/kg, 300 mg/kg, and 1000 mg/kg. This indicates that these Ag NPs may negatively impact the liver. In the bone marrow of either male or female rats, Ag NPs, on the other hand, did not cause any genetic damage. In addition, all tissues studied showed a dose-dependent increase in the concentration of Ag NPs. The amount of silver deposited in the kidneys varied according to gender. Over male kidneys, female kidneys had a twofold increase in the accumulation of silver. We did not see any differences in body weight or blood chemistry indicators between male and female rats after four weeks of nonstop exposure to Ag NPs for six hours per day on five days per week. Hepatic histology performed on Ag NP-treated rats showed cytoplasmic vacuolization and isolated necrosis, although this was not seen across the board. An animal study found that Ag NPs in the back muscles of rats implanted for 180 days caused granuloma development. As a result of the implants' small size and high surface area to volume ratio, macrophages surrounded and ringed them. The cytoplasm of macrophage cells contained many Ag NPs as well.

It has been shown in a separate study that zebrafish embryos treated with Ag NP had increased mortality and a delay in hatching at higher concentrations. Scientists used TEM to look for NPs in animal brains, hearts, yolk, and blood using electron-dispersive X-ray imaging. Several abnormalities were found in the bodies of the fish that had been exposed to AgNP, including distorted notochords and impaired blood flow, as well as an irregular heartbeat. These findings revealed that Ag NPs caused toxicity in zebrafish embryos dose-dependently. We discovered that mice's caudate, prefrontal cortex, and hippocampus areas showed significant oxidative stress and antioxidant defense array modifications at dosages of 100–1000 mg/kg of 25 nm silver NPs. When we applied Ag NPs to the frontal brain of mice, we saw an increase in ROS generation and gene expression alterations (25 nm).

Neurodegeneration and cell death were exacerbated by decreased glutathione peroxidase gene expression. As a result of this study's findings, researchers believe that neurotoxicity is caused by a change in gene expression and an increase in oxidative stress caused by free radicals. The reproductive capacity of C. elegant is decreased, and the gene expression of sod3 and daf12 is increased; this may have direct relevance to the inability of this species to reproduce due to exposure to Ag NP.

Copper NPs

Biocide research has focused on using copper nanoparticles (Cu NP) as an alternative to antibiotic therapy and a nanocomposite coating for microorganism growth inhibition. The histological study of mice exposed to Cu NPs after ingestion revealed significant damage to their kidneys, liver, and spleen. For mice, the LD50 of Cu NP (23.5nm) was 413 mg/kg, comparable to Zn powder's mild toxicity. On the other hand, cu microparticles (17 m) had LD50 values of >5000 mg/kg and had no harmful effects.

It was also shown that the mice exposed to copper NPs had renal tubular necrosis, glomerulitis, and renal tubule degeneration. However, the animals subjected to copper microparticles did not. According to Meng *et al.*, the deposition of Cu NPs in renal tissues is more effective than the deposition of micro-sized particles. According to the researchers, Cu NPs interacted with stomach juices and generated cupric ions in the kidneys. Cu NPs were acutely toxic to zebrafish, causing a decrease in the activity of the gill Na+/K+-ATPase. There was also an improvement in alanine aminotransferase (ALAT) levels after treatment with Cu

NP. After treatment with Cu NP, the gill lamellae disintegrated in a dose-dependent manner. This process was characterized by the proliferation of epithelial cells and the edema of primary and secondary gill filaments. RT–PCR results demonstrated greater gene expression alterations in zebrafish exposed to Cu NPs than in fish exposed to $CuSO_4$. Cluster analysis of these gene microarrays revealed an incredibly varied transcriptional response to Cu NP.

Titanium Dioxide NPs

Anatase (7–10 nm), rutile (15–20 nm), and nanotubes (10–15 nm diameters, 70–150 nm length) are three of the most often used nanostructures for TiO_2 synthesis, as well as rods and other geometries. TiO_2 is one of the most frequently created nanomaterials. TiO_2 has been demonstrated to have inflammatory and ROS-inducing effects on various cell and tissue types.

There are several uses for TiO_2 NPs, from paint to sunscreen to skin care products. Since many early experiments failed to take into consideration the different sizes and crystal structures of TiO_2, determining the underlying physical and chemical properties has proven to be difficult (anatase and rutile). It was determined by Braydich-Stolle *et al.* whether the crystal structure of TiO_2 NPs (with identical primary diameters) affected TiO_2's toxicity in the mouse keratinocyte cell line HEL-30. A 29-nm TiO_2 NP with an unknown crystal structure was able to influence inflammation and macrophage chemotactic responses in the lungs of rats more effectively than bigger 250-nm TiO_2 NPs. A lung epithelial cell line was likewise damaged by oxidative processes when exposed to TiO_2 NPs of varying sizes and compositions.

Researchers found anatase TiO_2 NPs produced apoptosis in all cells independent of size, unlike rutile TiO_2 NPs, which caused necrosis by producing ROS. According to Sayes *et al.*, anatase TiO_2 was more harmful than rutile TiO_2 in health risks. Using the BEAS-2B human bronchial epithelial cell line, other researchers have tested the toxicity of TiO_2 NPs at various concentrations (5, 10, 20, and 40 g/mL). We found that oxidative stress stimulated the expression of oxidative stress-related genes, including heme oxygenase 1, thioredoxin reductase, glutathione-S-transferase, catalase, and hypoxia-induced gene, as well as cell death and an increase in ROS. We also found a decrease in GSH and reduced glutathione (GSH).

There was also an increase in the expression of other inflammatory genes, including TNF-α, C-X-C motif ligand 2, and IL-8, all triggered by an external signaling pathway. Nano-sized TiO_2 rods and dots, however, elicited inflammatory responses in rats comparable to those caused by bigger TiO_2 particles in the lungs, according to an *in vivo* investigation. The capacity of macrophages to phagocytose other particles was dramatically diminished when exposed to nanometer-sized carbon black and titanium dioxide particles. Phagocytosis was significantly reduced in ultrafine particles compared to macro-sized particles.

In order to test TiO_2's low toxicity, Wang *et al.* gave it to mice at a level of 5 g/kg body weight. According to OECD guideline no. 420, the drug was given as a single oral dose. Hepatotoxicity (raised BUN level), pathological alterations in the kidneys, and TiO_2 accumulation in lung tissues were identified, suggesting that after being absorbed in the gastrointestinal tract, the NP might be transferred to other organs and tissues. Hydropic degeneration around major veins and patchy necrosis of liver cells were seen in the research of women exposed to 25- and 80-nm TiO_2 NPs. TiO_2 NPs immunogenicity was predicted in an alternate animal model employing a human autologous modular immune *in vitro* construct (MIMIC). Dendritic cell maturation and co-stimulatory molecule expression were enhanced by treatment with TiO_2 NPs in the MIMIC system. Compared to dendritic cells treated with micrometer-sized (>1 m) TiO_2, these treatments effectively primed CD4-T cells for activation and proliferation, a response similar to an in-vivo inflammation.

Cerium Oxide NPs

Polishing, computer chip production, and diesel exhaust emissions reduction additives benefit from using cerium oxide (CeO_2), a nano-sized substance. By activating Caspase 3, these NPs induced cell death by condensing the DNA, activating ROS production, and leading to apoptosis.

In cultured human lung epithelial cells, NPs with varying CeO_2 (15, 25, 30, 45 nm) sizes induced toxicity, increased ROS, decreased GSH, and activated genes associated with oxidative stress (such as heme oxygenase-1, catalase, glutathione-Stransferase, and thioredoxin reductase). As an alternative to this, additional studies have shown that ceria nanostructures have a high degree of biocompatibility, shield normal cells from radioactivity while doing little to protect tumor cells, and slow the pace of retinal degeneration produced by intracellular peroxidases. Nanoceria's ability to scavenge superoxides and operate as a catalyst may explain this apparent discrepancy. Cell-to-cell contact decreased, the mean total lamellar body volume per cell decreased, and 8-oxo guanine-positive cells increased after 10–30 minutes of exposure to flame spray-produced CeO NPs. NP-induced oxidative damage may lead to the release of surfactants as a defensive strategy.

Silicon and Silica NPs

To demonstrate that silicon microparticles are not damaging to cells, we observed intact mitotic trafficking of vesicles holding silicon microparticles in endothelial cells after they had been consumed. After the microparticles were digested, the presence of gold or iron oxide NPs in the porous matrix did not have any effect on the vitality of endothelial cells. For biomedical applications, silicon microparticles should facilitate the mitotic sorting of endosomes. NPs of amorphous SiO_2 were studied for their uptake, distribution, and cytotoxicity in mouse keratinocytes (HEL-30). Four concentrations of 30nm, 48nm, and 535nm SiO_2 NPs inhomogeneous solutions of the average size distribution (SiO_2 NPs) were tested for uptake and biochemical alterations after 24 hours of exposure. According to TEM findings, cells ingested silica of various sizes and stored it in the cytoplasm. LDH leakage was dose- and size-dependent when exposed to NPs with a size range of 30 to 48 nm. Both 118 and 535 nm particles were found to have no LDH leakage.

Mitochondria viability assay (MTT) results showed that the 30 and 48-nanometer particles were much more hazardous than the 535-nanometer particles at high doses (100 mg/mL). Further research was done to determine if SiO_2 toxicity is linked to cellular GSH depletion and mitochondrial membrane potential depletion. Cells' redox potential (GSH) dropped considerably after exposure to 30 nm NP at doses of 50, 100, and 200 g/mL. SiO_2 NPs with a size greater than 30 nm did not affect the levels of GSH. There was not a detectable difference in ROS generation between the control cells and the cells that had not been exposed to the agent. This study found that SiO_2 NPs smaller than 100 nanometers (nm) caused toxicity, indicating a vital role of particle size in producing biological consequences.

After 24 hours of exposure to 100 nm SiO_2 spheres at a 26.7 g/mL concentration, Brown *et al.* found that LDH leakage was only 3%. Thibodeau *et al.* investigated the role of SiO_2 in alveolar macrophage apoptosis in the study of pulmonary fibrosis in mice. Activation of Caspase 3 and 9 was seen in cells exposed to silica, as was depolarization of the mitochondria. However, it is not apparent what role ROS had in their inquiry. For four weeks, Kim *et al.* employed silica-coated MNPs to treat mice. Nearly all of the organs examined had NPs. Most of the NPs were absorbed by the liver and distributed throughout the body (*e.g.,* heart, kidney, spleen, and lungs). To their surprise, they discovered that NPs (50 nm) could cross through the blood-brain barrier and blood-testis barrier with no apparent injury. These findings imply that MNPs have the biological features necessary to serve as vectors for transporting genes and drugs [27].

Carbon-based NPs Toxicity

Single-walled Carbon Nanotubes

At high quantities, acid-functionalized SWCNTs (af-SWCNTs) were cytotoxic. Another research discovered that macrophages ingested -SWCNTs, subsequently identified in lysosomes, damaged mitochondrial function and inhibited phagocytic activity.

Multiwalled Carbon Nanotubes

The data on MWCNT-induced cellular toxicity is inconclusive. One study examined the toxicity of two different MWCNT types and found that caspase 3/7 activation did not cause apoptosis in RAW264.7 cells. Another study, however, found that MWCNTs that had been acid-treated (acid-MWCNTs) or functionalized with taurine (tau-MWCNTs) reduced cellular phagocytosis and increased cell death.

Depending on the type of macrophage, MWCNTs have different effects. BMDCs from mice died when exposed to MWCNTs ranging from 3 to 30 g/ml. However, RAW264.7 cells were unaffected by the same concentration of MWCNTs, even at 300 g/ml.

Fullerene

Fullerene (C60) did not affect alveolar macrophages in an MTT assay. On the other hand, human macrophages were not sensitive to C60, and they did not cause any inflammation in the cells.

Carbon Black NPs

Fe/CNPs (core/shell iron/carbon nanoparticles) may be used for MRI and medication delivery. Fe/CNPs (*e.g.,* pyrrolidone, primary amine, and alkyl alcohol) were applied to human embryonic kidneys (HEK293) and human cervical cancer cells (C33A). In alveolar macrophages and RAW264.7 cells, CB NPs promoted LDH leakage and IL-1 release. Acryl acid-functionalized Fe/CNPs were shown to have apoptosis-inducing characteristics independent of ROS production.

Nano Graphite

RAW 264.7 macrophages were shown to die from apoptosis and necrosis when exposed to pristine graphene. NG, CNTs, and CB toxicity in RAW264.7 cells were compared by measuring the release of extracellular LDH. Only those groups that received the highest doses of CNT and NG showed remarkable LDH release, and NG was more harmful to cells than CNTs. The increasing use of nanomaterials has resulted in various derivatives, some of which are harmful to cells. Graphene oxides (GO) were toxic to mouse peritoneal macrophages by Wan *et al.*

Single-walled Carbon Nano Horns

Carbon NPs are cytotoxic at low-uptake settings in a variety of studies. SWCNHs at 0.3 mg/ml triggered cell death in RAW264.7 cells, which included both apoptotic and necrotic processes, in a study reported in the Journal of Immunology.

SIZE EFFECT

Experimental evidence shows that, in comparison to bigger particles, smaller NPs have better cellular absorption and more cytotoxic effects. Silver NPs of a 20-nanometer diameter, for example, might enter the cell's interior and cause damage. NPs larger than 100 nm in diameter, on the other hand, cannot penetrate the cells.

A reduction in ATP generation and effects on other processes within the cell were observed regardless of the magnitude of either NP's reactive oxygen species (ROS) production. Occasionally, a significant degree of endocytosis efficiency has been recorded for 50 nm NPs. This is connected to the same vesicle size necessary for viral entrance at the first cell cycle stage for numerous viruses. The most effective strategies for penetrating H1299 human lung cancer cells and NCI-H292 pulmonary epithelial cells through pinocytosis include silica NPs and transferrin-coated gold NPs, respectively. Rather than a passive technique, endocytosis was the most efficient form of NP entry into HeLa cells. However, it was not known how the endocytosis occurred. NPs can interact with the lipid bilayer depending on their size. Pore-forming NPs in the lipid bilayer may severely distort it between 1.2 and 22 nm. However, the bilayer is no longer affected by NPs outside this range. This information might be useful for a single protein-NP attachment for bio-detection. Dendrimers' capacity to make membrane holes improved as their size rose from 4 to 8 nm, according to Leroueil *et al.* Dendrimer interactions with bilayers have shown that larger dendrimers can more efficiently cause pore formation. For bilayer pore growth, a perfect NP curvature is required. However, this depends on the cell type and NPs used. LDL stability is reduced by carbon nanoparticles (CNPs). The hydrophobicity of the carbon NPs in the bilayer's tail region affects the amount of stress needed to tear it.

By breaking the tail packing and producing undulating surfaces with free volume, the carbon NPs may reduce rupture stress and stimulate pore development inside bilayers. Reinforcing the tail packing of NPs with bigger sizes (between 0.72 and 1.2 nm) increases membrane rupture tension and prevents the creation of pores. Dendritic poly amidoamine (PAMAM) and dendrimers with a diameter of 2.32 nm might also be used to explain other observations, such as inhibition of pore development and stimulation of pore creation.

The aggregation behavior of NPs, when they interact with membranes, is also influenced by their NP size. According to the results of this experiment, the hydrophobic NPs measuring 2.5 nm gathered together to form clusters. In comparison, the 4 nm NPs produced chains of pearl-like structures. Many internalization mechanisms were found for NPs as large as six nanometers (nm), no longer membrane-bound NPs (5-9 nm). Since then, it has been simulated that the pearl-like chain aggregation occurs for longer NPs (7.5 nm).

NP aggregates have less impact on the membrane than a single NP of the same size. According to the findings of the researchers, smaller NPs had a larger curvature in the aggregates, which made them "visible" to individual lipids that were located nearby. DOPC liposomes could not swallow the smaller silica NPs (15-20 nm) in another investigation because they were too big (30-200 nm). Adhesion energy exceeded bilayer bending energy at a critical particle radius (CPR), attributed to a critical particle radius. In contrast to smaller NPs, larger NPs are more likely to be internalized into the liposome when smaller than this radius. The surface area of an NP is another factor to consider when determining its size. HSPG, a recently discovered protein that can trigger endocytosis, has the potential to pack more targeting molecules into larger NPs when operating in biological environments. This would increase the likelihood of contacting a receptor on the cell membrane, leading to endocytosis [28].

NP SHAPE EFFECTS

NPs that are not round, such as rods and tubes, may be easily produced. Their distinctive properties, derived from their shape, are increasingly being exploited in industrial applications. Furthermore, sphere cylindrical (SC) NPs are more efficient at translocating into the cell's membrane than other NP shapes (*e.g.,* sphere cylindrical).

Since they can better cross biological membranes and are used in composites to improve mechanical qualities, carbon nanotubes (CNTs) are a hot topic in medical research. Their small sizes and huge aspect ratios have been hypothesized to be responsible for their ability to penetrate membranes. The force needed to push a CNT through a lipid bilayer rose from 14 to 50 A.

CNT cellular entrance may be passive or endocytotic, depending on the particle diameter. Smaller sizes have been discovered to be able to pass through membranes. When the diameter is increased, endocytosis becomes the most energy-efficient method of transport. The same amount of force was needed to pass through a membrane for ellipsoids, rods, and disc-shaped NPs of similar size, according to the authors of this study. In order to differentiate between two separate methods of membrane penetration, studies were conducted using rod-like particles with varying aspect ratios and tip curvatures. Because they have a smaller aspect ratio and a flatter tip, they exert pressure on the membrane, which causes it to become invaginated. Particles having greater aspect ratios and more curled tips rotated such that they could establish complete contact with the membrane. This rotation was carried out until the membrane entirely covered the particle. After reaching a critical NP volume, higher aspect ratio rods experienced reduced energy bathers for membrane translocation. Other investigations have shown that NP uptake is unaffected by the NP form. Cellular absorption of diverse types of NPs has not yet been beneficial.

An investigation by Nature Nanotechnology has shown an edge in breaching the membrane in terms of volume. However, this edge is insufficient to warrant a significant difference in performance between the two shapes. Furthermore, a spherical-like cluster formed by an aggregated form of a tat-peptide was more efficient at translocating through the membrane than the original a-helical peptide.

It has been shown that pore formation may be initiated by NPs with a size of more than 22 nm and surface roughness on a macroscopic scale that comes into direct contact with lipid bilayers. In studies of NP uptake into HeLa cells using a variety of geometries (cubic, cylindrical, and spherical), it was shown that nanorods, and not spheres, demonstrated the greatest uptake rate. Since NPs have been placed in the bilayer area of the membrane and allowed to scatter or be forced into the membrane with pressure in the simulations so far, this is a long cry from a biological environment. There have been several types of research that have attempted to replicate dynamic physiological circumstances in the laboratory. For example, strong pressure is applied to the blood arteries to maintain continuous circulation. When the capillary wall membrane was mimicked by imitating the blood vessel state, the nanorods and spheres migrated to and interacted with the membrane. Shear rates may change the number of NPs that stick to the surface. Model capillary branching could cause NP-induced pressure disruption, resulting in a larger proportion of the capillary wall-bound (SPC) than the capillary-bound (SC) spheres remaining. When the capillary wall membrane was mimicked by imitating the blood vessel state, the nanorods and spheres migrated to and interacted with the membrane. Shear rates may change the number of NPs that stick to the surface. It has been shown that gold NPs with a diameter of 50 nanometers may depart the circulatory system and congregate around organs such as the heart, spleen, liver, and malignant tumors.

NPs with a diameter of fewer than 50 nanometers, on the other hand, might reach tumor cells at a deeper level. Nano hexapods absorbed more cells than the nanorods and cages, lowering their cytotoxicity. These findings show that infrared radiation may be used therapeutically to heat tissue to a few centimeters of depth. A pushpin NP design was created using dissipative particle dynamics and the geometric effect, allowing just the pin tip to penetrate a membrane when pushed rather than the whole pinhead.

It was possible to remove the pin from its membrane after the pressure had been alleviated. If drug molecules could be connected to the pushpin locations in a controlled drug delivery system, there would be no NP absorption in cells [28].

EFFECTS OF NP SURFACE CHEMISTRY

The presence of ligands on the surface of an NP has been linked to membrane thinning, erosion, and the creation of pores. Compared to their longer multi-tailed counterparts, NPs with shorter single-tailed ligands caused more membrane disturbances. Membrane penetration rates are influenced by ligand distribution on NPs, it was observed. For hydrophobic spherical NPs, the ligand distribution was optimized at a ratio of 3.5:2.5 between the top and bottom halves of the NPs to achieve optimum translocation efficacy (80%). The hydrophobic NP was partially exposed to encourage hydrophobic integration into the bilayer at lower densities. Passive endocytosis may be used to investigate the impact of NP surface chemistry on RBC cellular entry. Researchers have discovered that uncoated QDs create membrane holes in RBCs, resulting in cytotoxicity. When thiol D-penicillamine ligands were used to functionalize the QD surface, the membrane became more flexible, preventing pore formation and harmful effects. In other words, NP translocation time was the longest between σ ratios of 2:4 and 4:1 when ligands were uniformly distributed across the membrane (a = 3). This section focuses on the hydrophobicity and surface charge of the NP surface coating and how this affects NP cell entry [28].

NP SURFACE HYDROPHOBICITY

Hydrophilic and hydrophobic NPs are present on the membrane's surface, respectively. These hydrophobic carbon spheres, also known as fullerenes (C60), have a certain number of carbon atoms. C60 aggregation and membrane mobility might be generated by as little as one nucleotide for every nine DOPC lipid molecules, with the C60 molecules being dispersed and concentrated in the hydrophobic DOPC tail region. Because of its "barrierless" technique of membrane integration, the energy barrier that C60 had to overcome to penetrate a bilayer was almost nonexistent. Fullerene cytotoxicity was greatly lowered by functionalizing the surface. When the hydroxyl functionalized fullerenes were applied to the bilayer surface, the cellular absorption rate was lowered by nine magnitudes. This is not the only use of graphene in the POPC bilayer. It has been shown that up to eight graphene layers may be accommodated simultaneously. Orienting the graphene edges vertically across the bilayer was made possible by making the graphene hydrophilic. For example, Graphene sheets' conductive qualities could be useful in bioelectronics and sensors.

Hydrophobic silica NPs, on the other hand, have recently been found to cause bilayer curvature at lower temperatures than hydrophilic silica NPs. This finding is consistent with previous studies on DOPE. Hydrophilic NPs enhanced the lipid density around the hydrophobic surface of the DPPC membrane. For example, in an MD simulation, Rcynwar *et al.* observed that hydrophilic sections of NPs interacted with the bilayer. In contrast, hydrophobic groups exposed to the bilayer attracted one another, forming aggregate. A bilayer was deformed, and curvature was introduced into it as it aggregated. This helped to promote vesicle formation. However, the vesicle was unable to develop as it passed. The lipid density around semi-hydrophobic NPs' surface decreased when they penetrated the membrane without forming their lipid coating.

For hydrophilic NPs, AFM studies indicated they were more likely to develop a surfactant coating during bilayer translocation. Surface groups on NPs with many ligands have influenced their interactions with membranes. It is easier for gold NPs with striations to penetrate cell membranes when the ligands are hydrophobic and anionic. These NPs, however, had a different final destination in the cell. Several other investigations have shown that NPs with surface ligands organized in striations were more able to enter cells than those without.

The decreased energy barrier for particle membrane penetration is to blame for the ligands' cooperative interactions with the membrane. Structured ligands, on the other hand, maybe stiffer and more capable of permeating the membrane-like cell-penetrating peptides (CPPs). To further explain the improved cell absorption of NP with hydrophobic-hydrophilic domain strips. Small enough stripes may partly block off nearby voids and prevent quick NP aggregation when the solution is *in situ*. To facilitate their movement and incorporation into the membrane, NPs were kept somewhat unstable in water. The hydrophobicity of the NP's surface is likewise closely linked to its size.

When the size of an NP is reduced, it becomes more difficult for water molecules to form a hydrophobic layer on its surface. In water, bigger aggregates are more stable; hence smaller NPs tend to collect more rapidly after aggregating. If the NPs were reduced to the size of an anion, water could once again create a hydration shell around the ion (*e.g.,* Cl). NPs coated with numerous ligands may better integrate into membranes because of the higher mobility of ligands. As a result, the hydrophobic ligands and the positively charged ligands can easily separate into smaller NPs because of their greater curvature. The hydrophobic ligands are only present when the NP crosses through the membrane in the phobic area. At the same time, the charge ligands stay at the bilayer interface. This was also the case for rigid ligands that can accept this "snorkeling" process by rotating the ligands. Molecule membrane translocation can potentially be aided by hydrophobic NPs. It was discovered that hydrophobic NPs' affinity for membranes increased their translocation rates when associated with the surface of CPPs. The hydrophobicity of an NP may be controlled to hinder or assist membrane penetration, as seen in this example. By working together in six groups, CPPs can traverse a membrane naturally. Coating CPPs lower energy bather for membrane translocation with C60. The higher energy bather tethered to the NP hindered CPP penetration into hydrophilic NP [28].

EFFECTS OF NP SURFACE CHARGE

Due to their modest negative charge, mammalian membranes and positively charged NPs may interact electrostatically because the energy bather for membrane penetration is lowered. There is a more damaging reaction with positively charged NPs than with neutral or negatively charged NPs. Gold NPs with higher positive charge densities, for example, caused more damage to mammalian membranes. Different membrane penetration mechanisms have been proposed based on the NP surface charge density. The NPs formed holes in the bilayer when the surface charge coverage was less than 20%. Endocytosis was not detected until surface charge coverage was near 100 percent or greater, even when NPs were 40 percent or more covered by surface charge. These results imply that the surface charge density of an NP may affect its capacity to enter membranes and the method by which it does so. Anionic NPs may form more ordered areas in the bilayer, but in another work using DPPC bilayers, uncharged NPs had no detectable effect on the bilayer. Because of the electrochemical imbalance between the cytosol and extracellular environment that drives certain membrane activities and permits signal transmission, cationic NPs that connect to negatively charged proteins on the membrane alter the transmembrane potential. To increase membrane tension and create a hole in the cell membrane, a large amount of electrochemical potential must be drained. If transmembrane potential had not been abolished, endocytosis would have been the NPs' exclusive permeation mode. They could move through the membrane and be endocytosed once the restored transmembrane potential is restored [28].

AMPHIPHILIC PEPTIDES: RELEVANCE AND IMPLICATIONS TO NP TOXICITY

Amphiphilic peptides can disrupt membranes in various ways, including by altering their size and charge and their surface hydrophobicity and membrane composition. Because of their antibacterial properties and the fact that they are produced in the body, these endogenous peptides may have been selected for some reason.

As well as being useful in the medical field, these peptides may be used to purify water, fight cancer, and reduce obesity and inflammation in humans. Thirty-two amino acids make up an average-sized peptide. There is a net charge of +3.2 on average, making them amphiphilic. Disruptive membrane peptides' interactions with cell membranes are influenced by NP-like physical characteristics.

There are both opportunities and problems in fully digesting the positive effects of disruptive membrane peptides and correlating these findings with NP toxicity, especially cell membrane interactions and NP toxicity. As a result, we suggest that future research explore employing these peptides as model NPs due to their well-defined features. For information on these peptides' characteristics and prospective applications, we reference various in-depth research [28].

CONCLUSION

The primary drawback to employing NPs in medicine is the possibility of toxicity. As a result, research should be done on both the beneficial impacts of NP usage as well as the possibly unforeseen adverse effects of their impact on the human body. The distribution of NPs in the blood and lymphatic system, their capacity to enter nearly all cells, tissues, and organs, and their interactions with different macromolecules that change the structure of these macromolecules and impair intracellular and organ function are all factors that contribute to nanoparticle toxicity. Physical and chemical characteristics of NPs, such as size, shape, electric charge, and chemical makeup of the core and shell, have a significant impact on how poisonous they are. Even at very poisonous and deadly quantities, many forms of NPs are not detected by the body's and cells' defensive mechanisms, limiting their decomposition and perhaps causing considerable NP buildup in organs and tissues. Nevertheless, there are already a number of methods for creating NPs that are less harmful than conventional NPs. Current methods for investigating NP toxicity enable both the accurate prediction of probable adverse effects at the body level as well as the examination of distinct toxicity routes and mechanisms at the molecular level. It follows that it is impossible to design NPs that have little to no negative effects or even no effects at all unless all of the NPs' qualitative and quantitative physical and chemical characteristics are systematically taken into account and unless an appropriate experimental model for calculating their impact on biological systems is available.

REFERENCES

[1]   C. Buzea, I.I. Pacheco, and K. Robbie, "Nanomaterials and nanoparticles: Sources and toxicity", *Biointerphases,* vol. 2, no. 4, pp. MR17-MR71, 2007.
[http://dx.doi.org/10.1116/1.2815690] [PMID: 20419892]

[2]   A. Gnach, T. Lipinski, A. Bednarkiewicz, J. Rybka, and J.A. Capobianco, "Upconverting nanoparticles: assessing the toxicity", *Chem. Soc. Rev.,* vol. 44, no. 6, pp. 1561-1584, 2015.
[http://dx.doi.org/10.1039/C4CS00177J] [PMID: 25176037]

[3]   M. Lippmann, "Effects of fiber characteristics on lung deposition, retention, and disease", *Environ. Health Perspect.,* vol. 88, pp. 311-317, 1990.
[http://dx.doi.org/10.1289/ehp.9088311] [PMID: 2272328]

[4]   J.R. Gurr, A.S.S. Wang, C.H. Chen, and K.Y. Jan, "Ultrafine titanium dioxide particles in the absence of photoactivation can induce oxidative damage to human bronchial epithelial cells", *Toxicology,* vol. 213, no. 1-2, pp. 66-73, 2005.
[http://dx.doi.org/10.1016/j.tox.2005.05.007] [PMID: 15970370]

[5]   G. Oberdörster, E. Oberdörster, and J. Oberdörster, "Nanotoxicology: an emerging discipline evolving from studies of ultrafine particles", *Environ. Health Perspect.,* vol. 113, no. 7, pp. 823-839, 2005.
[http://dx.doi.org/10.1289/ehp.7339] [PMID: 16002369]

[6]   C.M. Sayes, J.D. Fortner, W. Guo, D. Lyon, A.M. Boyd, K.D. Ausman, Y.J. Tao, B. Sitharaman, L.J. Wilson, J.B. Hughes, J.L. West, and V.L. Colvin, "The differential cytotoxicity of water-soluble fullerenes", *Nano Lett.,* vol. 4, no. 10, pp. 1881-1887, 2004.
[http://dx.doi.org/10.1021/nl0489586]

[7]   C.J. Johnston, J.N. Finkelstein, P. Mercer, N. Corson, R. Gelein, and G. Oberdörster, "Pulmonary effects induced by ultrafine PTFE particles", *Toxicol. Appl. Pharmacol.,* vol. 168, no. 3, pp. 208-215, 2000.
[http://dx.doi.org/10.1006/taap.2000.9037] [PMID: 11042093]

[8]   J.T. Buchman, N.V. Hudson-Smith, K.M. Landy, and C.L. Haynes, "Understanding nanoparticle toxicity mechanisms to inform redesign strategies to reduce environmental impact", *Acc. Chem. Res.,* vol. 52, no. 6, pp. 1632-1642, 2019.
[http://dx.doi.org/10.1021/acs.accounts.9b00053] [PMID: 31181913]

[9]   Z.V. Feng, I.L. Gunsolus, T.A. Qiu, K.R. Hurley, L.H. Nyberg, H. Frew, K.P. Johnson, A.M. Vartanian, L.M. Jacob, S.E. Lohse, M.D. Torelli, R.J. Hamers, C.J. Murphy, and C.L. Haynes, "Impacts of gold nanoparticle charge and ligand type on surface binding and toxicity to Gram-negative and Gram-positive bacteria", *Chem. Sci. (Camb.),* vol. 6, no. 9, pp. 5186-5196, 2015.
[http://dx.doi.org/10.1039/C5SC00792E] [PMID: 29449924]

[10]   K.H. Jacobson, I.L. Gunsolus, T.R. Kuech, J.M. Troiano, E.S. Melby, S.E. Lohse, D. Hu, W.B. Chrisler, C.J. Murphy, G. Orr, F.M. Geiger, C.L. Haynes, and J.A. Pedersen, "Lipopolysaccharide Density and Structure Govern the Extent and Distance of Nanoparticle Interaction with Actual and Model Bacterial Outer Membranes", *Environ. Sci. Technol.,* vol. 49, no. 17, pp. 10642-10650, 2015.
[http://dx.doi.org/10.1021/acs.est.5b01841] [PMID: 26207769]

[11]   L. Lai, S.J. Li, J. Feng, P. Mei, Z.H. Ren, Y.L. Chang, and Y. Liu, "Effects of Surface Charges on the Bactericide Activity of CdTe/ZnS Quantum Dots: A Cell Membrane Disruption Perspective", *Langmuir,* vol. 33, no. 9, pp. 2378-2386, 2017.
[http://dx.doi.org/10.1021/acs.langmuir.7b00173] [PMID: 28178781]

[12]   D.N. Williams, S. Pramanik, R.P. Brown, B. Zhi, E. McIntire, N.V. Hudson-Smith, C.L. Haynes, and Z. Rosenzweig, "Adverse interactions of luminescent semiconductor quantum dots with liposomes and

shewanella oneidensis", *ACS Appl. Nano Mater.,* vol. 1, no. 9, pp. 4788-4800, 2018.
[http://dx.doi.org/10.1021/acsanm.8b01000] [PMID: 30931431]

[13]   M.N. Hang, N.V. Hudson-Smith, P.L. Clement, Y. Zhang, C. Wang, C.L. Haynes, and R.J. Hamers, "Influence of Nanoparticle Morphology on Ion Release and Biological Impact of Nickel Manganese Cobalt Oxide (NMC) Complex Oxide Nanomaterials", *ACS Appl. Nano Mater.,* vol. 1, no. 4, pp. 1721-1730, 2018.
[http://dx.doi.org/10.1021/acsanm.8b00187]

[14]   S. Mahendra, H. Zhu, V.L. Colvin, and P.J. Alvarez, "Quantum dot weathering results in microbial toxicity", *Environ. Sci. Technol.,* vol. 42, no. 24, pp. 9424-9430, 2008.
[http://dx.doi.org/10.1021/es8023385] [PMID: 19174926]

[15]   M.J. Gallagher, J.T. Buchman, T.A. Qiu, B. Zhi, T.Y. Lyons, K.M. Landy, Z. Rosenzweig, C.L. Haynes, and D.H. Fairbrother, "Release, detection and toxicity of fragments generated during artificial accelerated weathering of CdSe/ZnS and CdSe quantum dot polymer composites", *Environ. Sci. Nano,* vol. 5, no. 7, pp. 1694-1710, 2018.
[http://dx.doi.org/10.1039/C8EN00249E]

[16]   J. Khalili Fard, S. Jafari, and M.A. Eghbal, "A review of molecular mechanisms involved in toxicity of nanoparticles", *Adv. Pharm. Bull.,* vol. 5, no. 4, pp. 447-454, 2015.
[http://dx.doi.org/10.15171/apb.2015.061] [PMID: 26819915]

[17]   C. Outline, Oxidative Stress.*Biochemical Ecotoxicology* Elsevier: Amsterdam, 2014, pp. 103-115.
[http://dx.doi.org/10.1016/B978-0-12-411604-7.00006-4]

[18]   Y. Li, W. Zhang, J. Niu, and Y. Chen, "Mechanism of photogenerated reactive oxygen species and correlation with the antibacterial properties of engineered metal-oxide nanoparticles", *ACS Nano,* vol. 6, no. 6, pp. 5164-5173, 2012.
[http://dx.doi.org/10.1021/nn300934k] [PMID: 22587225]

[19]   G.A. Domínguez, M.D. Torelli, J.T. Buchman, C.L. Haynes, R.J. Hamers, and R.D. Klaper, "Size dependent oxidative stress response of the gut of Daphnia magna to functionalized nanodiamond particles", *Environ. Res.,* vol. 167, no. March, pp. 267-275, 2018.
[http://dx.doi.org/10.1016/j.envres.2018.07.024] [PMID: 30077134]

[20]   H. Chen, H. Yoshioka, G.S. Kim, J.E. Jung, N. Okami, H. Sakata, C.M. Maier, P. Narasimhan, C.E. Goeders, and P.H. Chan, "Oxidative stress in ischemic brain damage: mechanisms of cell death and potential molecular targets for neuroprotection", *Antioxid. Redox Signal.,* vol. 14, no. 8, pp. 1505-1517, 2011.
[http://dx.doi.org/10.1089/ars.2010.3576] [PMID: 20812869]

[21]   J. Fischer, M.H. Prosenc, M. Wolff, N. Hort, R. Willumeit, and F. Feyerabend, "Interference of magnesium corrosion with tetrazolium-based cytotoxicity assays", *Acta Biomater.,* vol. 6, no. 5, pp. 1813-1823, 2010.
[http://dx.doi.org/10.1016/j.actbio.2009.10.020] [PMID: 19837191]

[22]   V. Rabolli, L.C.J. Thomassen, C. Princen, D. Napierska, L. Gonzalez, M. Kirsch-Volders, P.H. Hoet, F. Huaux, C.E.A. Kirschhock, J.A. Martens, and D. Lison, "Influence of size, surface area and microporosity on the in vitro cytotoxic activity of amorphous silica nanoparticles in different cell types", *Nanotoxicology,* vol. 4, no. 3, pp. 307-318, 2010.
[http://dx.doi.org/10.3109/17435390.2010.482749] [PMID: 20795912]

[23]   Ü. Kumbıçak, T. Çavaş, N. Çinkılıç, Z. Kumbıçak, Ö. Vatan, and D. Yılmaz, "Evaluation of *in vitro* cytotoxicity and genotoxicity of copper–zinc alloy nanoparticles in human lung epithelial cells", *Food Chem. Toxicol.,* vol. 73, pp. 105-112, 2014.
[http://dx.doi.org/10.1016/j.fct.2014.07.040] [PMID: 25116682]

[24]   G. Fotakis, and J.A. Timbrell, "In vitro cytotoxicity assays: Comparison of LDH, neutral red, MTT and protein assay in hepatoma cell lines following exposure to cadmium chloride", *Toxicol. Lett.,* vol. 160, no. 2, pp. 171-177, 2006.

[http://dx.doi.org/10.1016/j.toxlet.2005.07.001] [PMID: 16111842]

[25] ho Lee Gang, Kim Tae-Jeong, and Chang Yongmin, Characterization*Ultrasmall Lanthanide Oxide Nanoparticles for Biomedical Imaging and Therapy*, 2014, pp. 43-67.
[http://dx.doi.org/10.1533/9780081000694.43]

[26] *Genotoxicity Hazard Identification and Characterization*. Second edition., 2020.

[27] A. M. Schrand, M. F. Rahman, S. M. Hussain, J. J. Schlager, D. A. Smith, and A. F. Syed, "Metal-based nanoparticles and their toxicity assessment", *Wiley Interdiscip Rev Nanomed Nanobiotechnol.*, vol. 2, no. 5, pp. 544-68, 2010.
[http://dx.doi.org/10.1002/wnan.103] [PMID: 20681021]

[28] C.M. Beddoes, C.P. Case, and W.H. Briscoe, "Understanding nanoparticle cellular entry: A physicochemical perspective", *Adv. Colloid Interface Sci.,* vol. 218, pp. 48-68, 2015.
[http://dx.doi.org/10.1016/j.cis.2015.01.007] [PMID: 25708746]

CHAPTER 6

Nanoparticles in Environmental Pollution Remediation of Xenobiotics

Abstract: Environmental deterioration is currently a major problem for both emerging and wealthy nations. Extensive industrialization and intensive agricultural activity are the main causes of land, water, and air contamination. There are numerous conventional treatments for various environmental contaminants, but each has drawbacks. As a result, a different approach is necessary, one that is efficient, less harmful, and produces better results. In terms of cleaning up the environment, nanomaterials have garnered much interest. Nanomaterials outperform more traditional methods for environmental remediation due to their enormous surface area and strong reactivity. For particular applications, they can be altered to include new functionalities. Nanoscale materials can be very reactive due to the high surface-area-to-volume ratio and a greater number of reactive sites. These traits enable greater contaminant interaction, which prompts a rapid decrease in pollutant concentration. In order to remove toxins from diverse environmental media (*e.g.,* soil, water, and air), environmental remediation primarily uses various methods.

Keywords: Environmental remediation, Pollution, Xenobiotics.

INTRODUCTION

The world is on the verge of a severe environmental catastrophe that will bankrupt us. The environment as it exists now is constantly decaying. Global environmental problems are escalating, and we must act as though there is an emergency in our world. We must adopt a fresh outlook and confront disasters with new ideas and tactics and our full knowledge and seriousness in advance. Nature takes millions of years to clean up pollution in the air, water, and soil. The two main sources of most environmental pollution are industry and automotive exhaust emissions. Humanity is at risk from air pollutants such as $NO(x)$, SO_2, highly reactive and dangerous organic chemicals, POPs including dioxins, and PAH (polycyclic aromatic hydrocarbons). Carbon monoxide (CO), when inhaled in large amounts, can cause rapid poisoning. According to the various levels of exposure, many heavy metals, including Pb, can cause immediate or chronic poisoning when ingested by living things. The above substances contribute to pulmonary conditions like COPD, asthma, bronchiolitis, malignancies, heart conditions,

Seyed Morteza Naghib and Hamid Reza Garshasbi

brain dysfunctions, and skin disorders. Condensation, flocculation, froth flotation, sand filtration, and AC adsorption are long-used techniques. These restrictions, however, include the ineffective scraping of metal ions, excessive energy input, and creation of non-recyclable compounds. Nanotechnology looks to be an emerging solution to these issues. Nanotechnology has the potential to contribute significantly to the creation of cleaner, greener technologies that have considerable positive effects on both the environment and human health. For their potential to offer solutions for pollution management and mitigation as well as to enhance the effectiveness of conventional environmental cleanup approaches, nanotechnology techniques are being researched.

By using less energy throughout the production and manufacturing processes, enabling products to be recycled after use, and creating and utilizing environmentally benign materials, nanotechnology can benefit the environment. Currently, nanotechnology shows much promise for solving sustainability problems, but we also need to consider any potential harm to the environment and human health.

NANO REMEDIATION

Using nanoparticles (NPs) to clean up contaminated water, soil, or air is known as nano-remediation. By adsorbing contaminants, accelerating the reaction, and lowering the hazardous valence to a stable metallic state, this innovative remediation technology has shown to be particularly successful at degrading toxins. The reaction is further sped up by the nanoscale, which creates a surface area with a greater optimal for adsorption. The carbon-based NPs CDs, GOs, and Carbon Nano Tubes (CNTs), as well as non-carbon-based NPs nZVI and zeolites, are some of the several nanoparticle agents employed in nano-remediation. However, it has been observed that the NPs employed in nano-remediation, such as GOs, rGOs, and CNTs, are harmful to human cells, particularly the lungs and breasts. The reactive surface with the exchangeable ions and the piercing size of the NPs are what cause toxicity. By adding functional surfactants to the reactive surfaces, these toxicities are typically reduced. The surface-functionalized NPs are, therefore, suitable for environmental applications, including cleanup [1].

SOURCES OF XENOBIOTICS AND POPS

All manufactured substances discharged into the environment have the potential to pose an environmental hazard. However, some may have a higher or lower dose than others. Excessive pollution is emitted from each source of xenobiotic contamination.

Xenobiotics can be released as a harmful component or a mixture of organic contaminants. Xenobiotics can use a single pollutant or a combination of pollutant substances, making environmental protection a major effort to be engaged in. The properties of xenobiotics have classed natural or manufactured releases as planned or accidental, direct or indirect. Pollutants emitted into the atmosphere by factories, which move to soil surfaces and waterways, are the primary source of most xenobiotics. It is necessary to classify xenobiotics' origins to limit the spread of potentially dangerous pollutants. According to the Stockholm Convention 2009, the transport potential should be computed as the sum of all discharges from a specific place. In huge quantities, people are exposed to POPs, halogenated xenobiotic chemicals, through various commodities and other non-point sources. The microbiome's food supply chain can be disrupted by these airborne contaminants, which are poisonous and highly bio accumulative. According to EPA regulations, the prevalence of illnesses generated by these poisons is high in marine and coastal ecosystems. Those who have adapted to contemporary conveniences by consuming synthetic chemicals are intentionally released into the environment. The four levels of POPs are (i) the most dangerous chemicals, (ii) the medium-level chemicals, (iii) the inadvertent release of chemicals during the production process, and (iv) the use of compounds that are currently being investigated.

The screening results from diverse communities at the international level will be used to take the necessary procedures for pollution management at the sites. The parameters were monitored regularly. Campaigns slowed down POP creation, use, and release. As non-degradable contaminants, they remain in the environment for a long time. Because organisms absorb these pollutants, their toxicity might fluctuate, or their metabolism can be disrupted.

ECOLOGICAL RISKS ASSOCIATED WITH XENOBIOTICS AND POPS

It has become a global issue since xenobiotics emitted by numerous businesses and agricultural activities have such a large impact on the ecosystem. This pollutant release has resulted in many diseases in food-producing animals, directly impacting human health. Organo-chlorinated insecticides, used in agricultural fields for decades, are the primary cause. It is impossible to eliminate these substances from the soil entirely. POPs are absorbed into the soil when wastewater is recycled. Therefore, it harms plant-microbe co-existence, damaging plants by changing their soil microbial populations. It also harms agricultural yields due to the imbalance in soil fertility. In other words, the roles of microbial biosynthetic pathways have been shifted, resulting in inadequate pollution cleanup. Ecological risk data predicted the adsorption of POPs in low-temperature zones. There is a chance that POPs will be transferred to marine habitats due to their migration to the high latitude atmospheres. Because of this, polar regions' marine areas have a higher accumulation than other regions' marine areas. As a result, residents in these regions face an increased risk of bioaccumulation because of their heavy reliance on seafood. Biological systems are always at risk from them. Nowadays, food, water, and commercial products are common sources of xenobiotic chemical exposure.

Toxicology and interactions with living beings; all cause the high risk for any substance. Ecological risk is affected by consumers' genetic features, the amount of time exposed to a certain chemical, and their dosage levels. Each chemical in the mixture should be counted and analyzed. This is the most important issue due to the microorganism's bioavailable nature for the xenobiotic substances in the mixture. The adsorption, degradation, and metabolism of xenobiotic chemicals differ depending on the kind of microorganism, making removal more difficult.

So, to combine environmental protection with the lowest costs, it is necessary to find effective risk assessment processes [2].

ROLE OF NANOTECHNOLOGY AND NANOMATERIALS FOR REMEDIATION OF POLLUTANTS

It may be said that research and development in nanotechnology are focused on subatomic and molecular size "materials. These scales defy all known principles of physics and chemistry. A material's physical properties change dramatically from the nanoscale to the macroscale. Carbon "nanotubes" for example, are a hundred times stronger than steel while being a sixth as light. In order to eliminate contaminants from the environment, the environmental remediation process includes a wide range of methods. Nanomaterials have a higher surface-to-volume ratio than macroscale materials, making them more biologically and chemically reactive. Because of their increased qualities and efficacy, they are particularly well-suited for these tasks. Science, the environment, industry, and medicine benefit from nanoscale materials. Nanoscale products with environmental remediation applications have experienced increased research and implementation in recent years. Nanomaterials have removed polluted soil and groundwater from hazardous waste sites. NPs may solve issues that standard approaches cannot because of their unique physical, chemical, and optical-electronic capabilities. It can devise novel methods for developing new methodologies, displacing existing tools, and creating new high-performing materials and chemicals while consuming minimal energy [3].

DIFFERENT TYPES OF NANOMATERIALS IN XENOBIOTIC TREATMENT

Nano-adsorbents

Nano adsorbents made of carbon, such as graphitized carbon and carbon nanotubes, are commonly used.

As a result of hydrophobic interactions, electrostatic interactions, hydrogen bonds, and covalent bonding, Organic contaminants can be absorbed by each form's adsorption sites, which are numerous due to the forms' flexibility and physical chemistry. Recently, high-energy sites have been formed in nanotubes by altering their surface chemistry. When porosity rises, hazardous chemical sorption sites may be removed using magnetic NPs. With the production of carbon dots in various ways, these magnetic carbon nanotubes can change the surface. The adsorption rate was relatively high, indicating a high degree of reusability. Sol-gel methods that use multiple-walled carbon nanotubes with negatively charged surfaces are designed to remove cationic pollutants. Core shells made of porous polymeric nano adsorbents coupled with magnetic NPs may extract contaminants from the environment. Hybrid structures were successfully constructed using this synthesis process to remove heavy metal ions with a broad pH range that are very absorbent.

Nano-filters

Filtration technology has reached new heights since the introduction of electrospinning. By lowering the ionic strength of the solution, nano filters lessen the abrasiveness of organic pollutants. Different natural and manufactured polymers, such as Polyvinyl chloride, polypropylene, and polyacrylonitrile, are used to create nano porous membranes. Work on nano-filtration focused on a pressure-driven approach that effectively eliminates components with molecular weights between 10 and 1 nm. It filters using hydrodynamics and membrane nanopores on the membrane surface. The efficacy of membrane charge, porosity, and surface concentration polarization is an important factor in the filtering process. Electro-spinner technology is used to construct high-quality, linked 3D membranes. Nanofibers are highly successful at rejecting one, two, or more valent ions. As a result, treating drinkable water to remove arsenic is the best option.

Nano-filtration effectively removes nearly all dissolved salts while consuming less energy and operating at a lower pressure than other filtering methods. Nano-filtration membranes are commercially available for the removal of arsenic compounds. Arsenic compounds can be removed from water using nano-filters made of highly charged polyamide. Physicochemical interactions between membranes and contaminants are commonly used to separate and eliminate pollutants; however, only nano-filtration is possible for trace components like pesticides. Nano-filters have a range of molecular weights and hydrophobicity properties that limit their removal ability for dissolved organic chemicals and uncharged insecticides. There were 11 different pesticides that the nano-filter membrane could not remove, including aromatic and hydrophobic pesticides. The pesticides' polarity modified the filtration capacity of the membrane. Because charged pesticides are close to membranes, the separation is successful. As a result, nonpolar membranes are used to make hydrophilic nanofillers to prevent organic contamination.

Nanofibers

To meet the minimal criteria for water purification, membrane filtration is essential since flocculation, sedimentation, coagulation, and activated carbon cannot remove organic impurities effectively. Polyvinyl fluoride, polypropylene, and polyacrylonitrile are polymers used in nanopore membrane synthesis. These nanofibers may effectively remove micropollutants from the wastewater. As opposed to tubes and particles, they have much looser bundles. Molecular propagation pathways play an important role in pesticide contamination adsorption. In order to polymerize atrazine herbicide nano-fibrous membranes, pyrrole is utilized. Toxic industrial pollutants can be treated with photocatalytic nanofibers manufactured from semiconducting materials.

Titanium dioxide and graphene-polymer composite nanofibers show a solid photocatalytic effect on the degradation of different dye compounds. When arsenic comes into contact with the membrane, it alters the membrane's ionic potential, reducing the concentration of sodium and calcium ions. When used in wastewater treatment, these nanofibers can effectively remove endocrinologically active micro-pollutants from the environment. Antibiotics (Ciprofloxacin and Bisphenol) are adsorbed onto polyacrylonitrile nanofibers generated by electrical spinning techniques. Reduced molecular dimensions improve total absorbance. PA6 nanofibers are coated with nanofibers polymerized to adsorb atrazine herbicide. Using bacterial cell bio polysaccharide cellulose is necessary to create extremely thin, multilayer nanofilms. It has been cyclodextrin-coated and is now being utilized for water cleaning. POPs like phenol, bisphenol A (BPA), and glyphosate are effective adsorbents in this film (2, 4- DCP). The product's remarkable reusability allowed it to display outstanding adsorption over a wide pH range.

Nanocomposites

The adsorption of contaminants has a greater impact on biopolymer nanocomposites than on micro-and monolithic agents. Integrating adjustable features, including electrical, mechanical, and magnet capabilities, new hybrid matrices have been constructed that effectively store pollutants and deliver harmful payloads. Nanocomposites have a higher recycling rate than other materials, making them an appealing pollutant removal medium. Co-precipitation, hydrothermal thermal deposition synthesis, sol-gel synthesis, microwave synthesis, chemical vapor deposition, surface modification, and energy-efficient ball milling are the only methods used to make nanocomposites.

Stability is an issue when using nanocomposites to remove xenobiotics, such as adsorbents or photocatalysts, although the benefits outweigh these drawbacks. Metal ions may be leached from the nanocomposites. Because of this, the commercial manufacture of nanocomposites is limited by environmental and production costs.

Graphene-based Nanocomposites

One of the most common uses of graphene is to remove inorganic pollutants and crude hydrocarbons from the environment. Depending on how it is adsorbed, it can perform various tasks. Carbon derivatives are developed to address the issue of activated carbon, which is difficult to remove and is a persistent organic contaminant. Many graphene nanocomposites have been developed as a sorption medium to remove aromatic contaminants. The BET (Brunauer–Emmett–Teller) test results demonstrate that nano porous graphene has higher phenanthrene and persistent pollutant adsorption potential. Their structures have delocalized - electron interactions with the aromatic rings of contaminants that interact with comparable stacking interactions in the pollutants. Cadmium, chromium, and lead are examples of cationic metallic pollutants affected by the functional oxygen-bearing groups on the surface of graphene. There is an exponential effect on the ability of cationic metallic pollutants to adsorb, including cadmium and chromium, and lead.

There are graphene layers in Fe_3O_4-rGO nanocomposites that help eliminate lead and arsenic. This nanocomposite has high recyclability, which helps remove rhodamine chemicals. Phenolic contaminants can be removed using graphene nanocomposites functionalized with amines. The adsorption capability of tetracycline contaminants in water was investigated for graphene nanocomposites functionalized with thiourea dioxide and iron (III) oxide. The hydrophobic interactions between tetracycline molecules and thiourea result in higher thiourea adsorption than magnetite.

Zhang and colleagues investigated the adsorption capability of the pesticide contaminant ametryn on a cellulose-graphene oxide adsorbent. Activated carbon was outperformed by iron functionalized with graphene oxide in the adsorption of fulvic acid. The adsorption rate was seven times lower than the $Fe_3O_4^-$ graphene oxide when the sample pH was around 5. Compared to graphene oxide nanocomposites, magnetite nanocomposites are more effective at adsorption.

Magnetic Nanocomposites

Bi_2O_4-coated iron (III) oxide nanorods efficiently degrade the persistent organic contaminant ibuprofen. Five times more efficient than conventional nanocomposite materials, this nanocomposite can be recycled magnetically and utilized repeatedly. Regarding adsorption efficiency ranges, activated carbon-coated magnetic NPs have been used to treat dyes in naked red nylon, nylon, blue, or chromazurol. Using magnetic particles and silica, self-assembled nanocomposites were created to remove arsenic compounds. A simple ball milling technique created TiO_2/graphene oxide/$CuFe_2O_4$ reusable nano propylene, enhancing photocatalytic properties. A photocatalyst was employed with UV light to remove 17 pesticide residues from the atmosphere. Polycarrageenine copolymers with polyvinyl acetate and Fe_3O_4 were tested to create nanocomposite hydrogels and their adsorption capacity using a crystal violet dye solution. According to these observations, as the content of carrageenin increases, so does the efficacy of adsorption in the crystal violet solution.

Polymer Clay Nanocomposites

A filler ingredient is combined with another compound (clay) to make polymer nanocomposites. This structure comprises natural or synthetic polymers that form a matrix. Chemical reactions at higher temperatures alter the polymer matrix, causing it to degrade. Polymers with great temperature resistance have been developed due to significant technological advancements. In the natural environment, these thermo-resistant polymers degrade xenobiotic components. The deterioration efficiency is determined by the degree to which the matrix fill has been encapsulated. Polystyrene matrix loaded with phenol, chlorophenol, and nitrophenol was strongly adsorbent. Four nitrophenols and copper were recovered from an alginate nanocomposite containing montmorillonite (MMT).

PDADMAC and Polymer Clay (PC) poly(4-vinylepyridincostyrene) composites effectively reduce the concentration of atrazine herbicide in MMT clay. A mixture of magnetite and bentonite is employed as an adsorbent to remove nitrofurazone from water to 50%. Adsorption of clopyralid in aqueous solutions was stimulated by sustaining temperature variations in the presence of MMT (SWy-2)–chitosan bio-nano composites (SW–CH). Because the nanocomposite preparations contained bio polysaccharides, the removal efficiency was increased. Calcium alginate beads nanocomposites covered with iron oxide NPs adsorb ortho nitrophenol compounds. The ideal pH for achieving a 96% clearance rate was 2. Adsorption rates of perlite beads covered with chitosan nanocomposites were greater at pH 7.0. Poly (4-vinyl pyridine-co-styrene) and poly diallyl dimethylammonium chloride with MMT nanocomposites efficiently eliminated trichlorophenol and trinitrophenol contaminants.

Nano Sponges

The removal of organic and a polar pollutant by cyclodextrin polymers occurs in sections of one trillion, as opposed to zeolites and activated CO, which are eliminated in ppm concentrations, according to research on the polymers. It is impossible to improve the cyclodextrin adsorption quality by recycling it more than twice because of its insolubility. Hydrophobic molecules can be trapped by cyclodextrin compounds, which feature interstitial pores in the host's interior cavities.

They can be used in many scientific endeavors, particularly in environmental studies. Chlorinated aromatic compounds like 4-chlorophenoxyacetic acid and 2,3,4,6-tetrachlorophenol may be removed from water using cyclodextrin nano sponges. When the pesticide sequestration is complete, it is possible to overlay magnetically designed nanoparticle-coated sponges to increase the polymer's characteristics. Nano sponges impregnated with polycyclic aromatic hydrocarbons, tri-halogen methane, monoaromatic hydrocarbons, and pesticide simazine have been proven more effective on porous ceramic membrane filters than other forms methods of removing these pollutants.

Nano sponges containing cyclodextrin derivatives (CDNS) were employed to remove rhodamine B from the environment. The adsorption capacity of CDNS changes with dye structure. The photocatalytic degradation of phenol was studied using beta cyclodextrin-nano sponge polyurethanes embedded in titanium dioxide/silver NPs. Cyclodextrin nano sponges may be combined with nano polymeric beads, dendrimers, nanomembranes, nanofibers, and carbon nanotubes to create composites that can be used to remove pollutants from water. To remove oil from water, a carbon nanotube-based synthetic nano sponge is utilized. Nano sponges make it simpler to recover and repurpose oil. When two organoclay polymers cross-link to create cyclodextrin, the triclopyr (3,5,6-trichloro-

2-pyridinyloxyacetic acid) is decomposed, and elimination efficiency is around 92%.

Nano-zeolites

There is much interest in using natural microporous materials like Zeolites to remove color impurities because of their ability to exchange ions. The ionic characteristics of zeolite have been improved, and its catalytic level altered. Waste has been degraded as a result of these efforts.

Due to specific properties, however, zeolite cannot be used as an adsorbent because of its low permeability, decreased adsorption capacity when humidity exposure, and difficulties in removing heavily suspended particles. Cadmium ions have been more effectively removed from water by electrospinning polyvinyl alcohol/zeolite nano adsorbents. Recent studies have shown that zeolites may be used to recover inorganic cations and hydrophobic organics.

Nano Sensors

The researchers first developed sensors for air and wastewater pollutants. Tiny photocatalytic devices using semiconductor nanostructures for pollution detection have been constructed on microscopic scales by integrating nanotechnology with microelectromechanical systems. Pesticide contaminants can be converted into innocuous chemicals like carbon dioxide, nitrogen, and water using manganese-doped zinc oxide nanowires in visible light. Using a solar simulator, a cylindrical photochemical reactor with aqueous TiO_2 dispersions was used to degrade (3,6-dichloro-2-methoxy benzoic acid). Titanium dioxide thin films degrade endosulfan after just 45 minutes of UV exposure. Semiconductor-based nano sensors may detect herbicide contaminants using soil samples as a starting point. After photodegradation, the contaminants are entirely absorbed and degraded.

Metal and Metal-oxide NPs

The environmental friendliness of metal NPs has sparked considerable interest. As metals and metal oxide NPs have many surface reaction sites, they are widely used to eliminate pesticide contaminants. Aside from absorption, they can also reduce the environmental impact of pollutants by turning them into less dangerous products. NPs size quantization characteristic is critical to the cleanup of the pollutant.

Graphene oxide NPs can interact with metal ions, making them reactive to arsenic metal ion contaminants and immobile during the cleanup. They contain an abundance of chemically bound oxygen on their surfaces.

Organophosphorus pesticides may be eliminated using nano metal oxide NPs and alumina. C-18-coated magnetic NPs remove nonpolar herbicides. Many organic dyes can be removed using zinc oxide NPs, and zinc oxide NPs may remove permethrin compounds using rapid reactive adsorption techniques completely. Iron oxide NPs have removed As (V) and As (III) from the environment. It has also been shown that zerovalent iron oxide NPs may be utilized to remove heavy metals from industrial waste streams, such as chromium and lead. Nano zerovalent iron also can dechlorinate halogenated insecticides, herbicides, radionuclides, and organic compounds. Atrazine, diazinon, and diuron are all nitroaromatic substances that may be converted into their amines by nZVI compounds. When exposed to ultraviolet light, tin oxide NPs may act as photocatalysts. NPs that interact with pesticides excite hydroxyl radicals, which are then released. Reports show that titanium dioxide NPs can convert dicofol, a chlorinated pesticide, into less hazardous chlorine molecules; the degradation of most chlorinated pesticides will occur in a shorter time. Particle size, noble metal doping, active surface sites, and particle shape may all be modified to improve the adsorption capacity of nanoparticle surfaces. Nano alumina, mostly used to remove mercury contaminants, is another typical nanoparticle with great thermal stability. The antibacterial capabilities of silver NPs are meant to eliminate membrane fouling concerns. Hydrophilicity, permeability, homogeneity, separation time, and fouling resistance may all be affected by NPs. It is possible to safely remove metal contamination from soil and water using biogenic NPs [2].

NANOMATERIALS IN REMEDIATION OF AIR POLLUTION

To enhance air quality, nanotechnology may be applied in several ways. For gaseous processes, nano catalysts with a greater surface area can be used. Cars and industrial buildings emit dangerous fumes, and catalysts speed up chemical reactions that turn them into a harmless gas.

Disposing of filtered chemicals is challenging since air waste is not effectively removed from the ground. When it comes to air pollution, filters are a lifesaver. Porous design enables gas to pass through yet retains particles within the filter. The three factors that enable membrane treatment are an electrostatic charge created by the particle/filter design, inertia from moving gas, and contact with the surface. Filters have a major drawback: they generate a significant pressure drop, which requires much energy. As its holes vary from 1 to 10 nm, nano filters are much more efficient at removing bacteria and organic pollutants from water than normal filters. For separating CO_2 from other gases, carbon nanotube-based membranes are better suited. It is possible to capture gases using carbon nanotubes 100-fold faster than the present gas separation method. The outcome is that large-scale applications can be implemented with them. Traditional layers negatively correlate gas separation efficiency and flow through the corresponding volume, but carbon-based nanotube membranes cover this. There are several industries where nanomembrane technology can separate and purify gases and pollutants and prevent their release into the atmosphere. Nano sensors can detect and respond to physical changes on a nanometer scale.

For businesses, obnoxious and hazardous spills can be one of the most significant risks. Small concentrations of these gases have lately been detected using commercial-level sensors. This necessitates the development of faster and more accurate sensors thanks to recent technological advances.

Small, round, 3D NPs are utilized as sensors to detect the presence of toxins in an environment. This biosensor type can catch harmful gas molecules using single or many membrane nanotubes. A new type of nano sensor is smart dust. The goal is to create a series of ultralight nano computers that can be used as advanced sensors in the long run. Data may be sent wirelessly to the central server using silicon-based ultrafine particles. This type of sensor can transmit data at one kbps or more, which is critical. These nano sensors can be suspended in the air for long periods, fueled by solar energy. Smart dust can communicate climate, humidity, and pollutant levels up to 20 kilometers away. This provides the necessary circumstances for continuous pollution management in a specific area.

CARBON-BASED MATERIALS

Remediation, prevention, monitoring, and sensing are some of the advantages that nanotechnology has for lowering air pollution. Oxygen-containing functional groups are abundant on the surface of graphite oxide, and these groups can be regulated by adjusting the reaction temperature by including water. Ammonia gas sensors working at various temperatures have been made of this material [4]. A novel material with many qualities can be created by combining carbon-based NPs with other nanomaterials to create nanocomposites. The adsorption of SO_2, GO, and zirconium hydroxide/graphene composites has been used for environmental cleanup. Moreover, GO was utilized as a photocatalyst to degrade VOCs and was partially decreased *via* photoreduction when exposed to ultraviolet light [5].

PM 2.5 was also captured using a GO membrane with a significant surface area and constant pore structure. CNTs have been the subject of numerous investigations to improve their adsorption abilities. CNTs must be modified or coated using other reactive materials with the right functional groups or charges [6].

SILICA-BASED MATERIALS

Silica-based nanomaterials provide a wide range of advantages, including a large surface area, variable pore size, and easily changeable surfaces. They are remarkably flexible. Also, in recent years, interest in using these nanomaterials to clean up dirty air and remove contaminants from the gas phase has increased because of their capacity for catalysis and adsorption. The physicochemical properties of silica NPs may be improved through surface modification. For instance, silica NPs hydroxyl groups on the surface may help produce a variety of surface phenomena, such as gas adsorption and wetting. The method works well for creating brand-new catalysts and adsorbents. It was shown in one of the earliest investigations into the adsorbent capability of modified mesoporous silica that the inclusion of amine groups on the surface improved the efficiency of H_2S and CO_2 collection from natural gas. The scientists claim that the material is extremely effective in removing such gases because it swiftly eliminated up to 80% of the total H_2S (in 35 minutes) and CO_2 (in 30 minutes) [7].

Similarly to this, another study found that amino silicates may be able to remove CO_2 from the surrounding air. This finding raises the possibility that these substances can reduce climate change. These amine-modified silicates remove organic pollutants, including aldehydes and ketones, and CO_2. They might therefore be useful for eliminating contaminants from an industrial setting. On the other hand, lead (Pb) pollution of the atmosphere is a global environmental and health issue that is difficult to eradicate. To address this issue with the environment, Yang *et al.* [8] created silica NPs. Their silica NPs were able to eliminate ambient Pb from contaminated air based on their results. As pb and other heavy metals create industrial pollution, silica-based NPs may be useful environmental agents to fight this.

NANOMATERIALS IN REMEDIATION OF WATER POLLUTION METAL AND METAL-BASED NANOMATERIALS

Because of their high reactivity, photolytic properties, and adsorbent qualities resulting from their large surface area and affinity to various chemical groups, several types of metal oxide NPs, including iron oxide (Fe_2O_3/Fe_3O_4), zinc oxide (ZnO), and titanium dioxide (TiO_2), are used for water purification [9]. Despite their powerful adsorption capability and durability in suspension media, iron NPs have been used to neutralize colors in wastewater from the textile, paint, and paper industries. Recently, it has been demonstrated that methylene orange and methylene blue, two of the industrial dyes with the greatest negative effects on the environment and human health, are extremely successful at being absorbed by these NPs [10]. Magnetic iron oxide NPs and carbon were used to remove methyl orange and phenol, and the results showed that the nanocomposites have a greater contact with the color, with the amount of carbon being a key factor in the NPs' adsorbent activity [11]. In addition to dyes, another significant category of water contamination includes heavy metals like chromium (VI). Recent research suggests that organic acids, iron oxide, and zerovalent iron NPs may all contribute to reducing the environmental harm that chromium (VI) poses (such as citric acid). Titanium dioxide NPs (TNPs) are an effective alternative to recently discovered poisons like pharmaceuticals and a popular photocatalyst for removing micropollutants from water [12].

CARBON-BASED NANOMATERIALS

Examples of nano porous carbon-based materials with physicochemical properties suitable for the treatment of water to remove contaminants like heavy metals, fluorides, textile dyes, and pharmaceuticals include carbon nanotubes (CNTs), including multi-walled and single-walled nanotubes (MWCNTs), graphene, and its oxide. For instance, one study investigated the hexavalent chromium adsorption by MWCNTs in contaminated groundwater [13]. The authors examined the impact of pH and adsorbent concentration on adsorption efficiency. Their findings revealed that the adsorption decreased at pH levels greater than 7. Projects to remove gasoline from water have also used MWCNTs [14]. Due to the significant environmental risk that fluoride poses, various carbon-based alternatives have been used to defluorinate wastewater. In this respect, investigations on the ability of chemically and biologically reduced graphene oxide to remove fluoride show that the former had an 87% reduction while the latter had a 94% capability [15].

Similar to how activated carbon's low cost, large pore size, and high porosity have been extensively investigated in removing medicinal compounds. For instance, a study comparing the adsorption of sildenafil citrate and carbamazepine onto granular and powdered activated carbon was published in 2019 [16]. As a result of the larger surface area of powdered activated carbon, the results showed that 90%of the compounds were removed from the mixture in just 10 hours, as opposed to just 40% for granular activated carbon after 70 hours. It was also reported that testing the adsorption of caffeine, ibuprofen, and triclosan using powdered activated carbon revealed a significant pH effect [17].

NANOMATERIALS IN REMEDIATION OF SOIL

Humans and ecosystems are severely harmed by polluted water discharged by industry, untreated sewage, and synthetic toxins from homes. Pollutants that risk human health include heavy metals and organic chemicals. Innovative nanomaterial-based technologies have been developed thanks to nanotechnology.

When removing contaminants from the environment, NPs may be produced cheaper and with a bigger surface area per unit weight than conventional adsorption methods. Several different types of nanomaterial-based adsorbents have been investigated in this regard. These include zero- and one-dimensional NPs, nanofibers and tubes of just one dimension, and 2 and 3-dimensional nanosheets and flowers.

A wide range of nanomaterials, both carbon-based and non-carbon-based, can be used to aid with water purification. The interaction of the contaminant with the surface of the nanomaterial influences the segregation process. Furthermore, the permeable segregator unit may use NPs as a key manufacturing component of the ultimate filtering membrane or a thin film to improve the hydrophilicity and thermomechanical capabilities of the membrane. Zinc and titanium oxides, ceramic nanowires, polymeric membranes, carbon nanotubes, and submicron particles are among the materials used.

The carcinogenic properties of dye pollution make them lethal. It has been thousands of years since dyes were first employed to color textiles, paints, and pigments. Synthetic dyes are now mass-produced in industrial quantities in numbers approaching 0.1 million. The annual consumption of dyes is expected to be 1.6 million metric tons. About 15% of the total is lost in the process of usage. Chitosan sugar-coated magnetite NPs have good adsorption ability for crocein orange G and acid green 25.

Intriguingly, the adsorbent NPs can be easily recovered using a magnetic field. Methylene blue and bright green dyes containing 138 mg/g of Fe_3O_4/activated carbon NPs were successfully separated using these NPs. Vanadium NPs in electro-spun polyether sulfone nanofibers remove methylene blue dye from water. The nanofibers may adsorb cationic Methylene Blue molecules exist in an alkaline environment due to their low isoelectric point and large, highly hydroxylated surface area. Adsorption's main problem is that dye contaminants cannot be "eliminated" or "neutralized" to their original state. To gather color components, molecular phase shifts are used rather than ion exchange. Due to the difficulty of removing dye-based sewage and depositing it on the membrane, this capacity may be challenging to deploy. Promising oxidation strategies based on hydroxyl radicals (OH) released from textile effluent have been identified. The unpaired electrons in the hydroxyl radical make it a potent oxidizer of organic pollutants resistant to other oxidizing agents. It is now widely accepted that Photocatalysis, an improved oxidative remediation approach, is a viable way of destroying dyes (organic) and pesticides because of its low cost. Metal oxides with semiconductor characteristics, including zinc and titanium, have shown outstanding photodecomposition efficacy when employed to remove dye impurities.

Whenever exposed to light semiconductors with energy gaps in the visible light spectrum, industry-affiliated light photocatalysts exhibit low stability or efficiency. There are many wonderful and workable answers to these problems. There is still a big environmental problem with soil contamination and deterioration. The issue of waste disposal is growing more prevalent around the globe. Soil deterioration worldwide has a significant impact on food production and safety. Hence action must be taken immediately. Heavy metals, herbicides, and POPs (persistent organic pollutants) have been identified in the soil, exacerbating the situation. The danger of food chain poisoning is increased due to the biomagnification of these substances in the soil. The demand for food and land degradation issues significantly affect agricultural productivity.

Adsorption or immobilization is one of the most often utilized modalities of therapy. Removing metal contaminants from soils is easier with this type of remediation technology because it is practical, inexpensive, and environmentally friendly. Artificial carbon nanomaterials, such as carbon nanotubes, metal oxides (ferric oxide and titanium oxide), and other nanocomposites, immobilize soil pollutants. When absorbing and immobilizing heavy metals, ferric oxide NPs are an excellent example. Oily sewage is a concern for the local aquatic life and should be avoided.

Total Petroleum Hydrocarbon (TPH) removal by iron NPs was increased by 88.34%. This means that water treated with nanotechnology has fewer pollutants and hazardous compounds while also being able to remove heavy metals. In order to lessen the dangers of environmental pollutants, nanoscience can be used in conjunction with biological treatments. Industrial techniques backboned by NPs can efficiently decolorize organic dyes. Gold and silver nanoparticle catalysts work well in a two-step process to degrade dyes. First, electrons accumulate on the particle surface, and then the dye diffuses to cause a reduction reaction. Magnetic NPs can be used in a wide range of pollution control applications.

Abiotic reductive reactions can be used alone or with immobilization strategies to remediate redox-sensitive contaminants by directly reducing their mobility and toxicity in soil. Polyacrylamide-modified magnetite NPs can reduce soil erosion and the leaching of arsenate from the environment. There has been great interest in pollutant immobilization on-site as a practical and cost-effective remediation technique.

Full-form changes of NMs to discover a material with low cost, high efficiency and stability, and minimal environmental impact have been researched. When nanohydroxyapatite particles reduced metal concentrations in pore water, they successfully immobilized them.

In this investigation, sodium carboxymethyl cellulose-stabilized nZVI removed around 80% of the soil-bound Cr (VI). Pyrene could be reduced by adding nZVI to an infected soil sample that had already been treated with the herbicide. Aside from the limitations of early acidification, such as its high cost and detrimental influence on soil quality and microorganism quality, these activities are limited. Nanomaterials are also employed to break down organic pollutants in polluted soil using advanced oxidation processes (AOPs). When combined with soluble Fe (II), hydrogen peroxide or persulfate may form extremely reactive radicals (such as water radicals or sulfur dioxide radicals), which can be employed to degrade organic contaminants by oxidative degradation. Solid iron phases have been suggested as an alternative to soluble Fe (II) to minimize pH shifts during chemical oxidation. When combined with chelating ligands, iron particles provide an added benefit to the process. Using green NPs to combat environmental pollutants has many potential applications, including treating hazardous metal ions in limited areas. Remediation and cleansing of waste sites can benefit from using these products. Ferrous NP can cleanse water and soil contaminated with heavy metals and fertilize environmentally friendly soil. For solar energy applications, siliceous material produced by bacteria has been used in optical coatings [3].

METAL AND METAL-BASED NANOMATERIALS

Nanoscale zerovalent iron (nZVI) is an electron donor with a low reduction potential. nZVI is one of the most commonly used compounds in pilot studies because it allows for the elimination of polychlorinated biphenyls, organochlorine pesticides, and chlorinated organic solvents *via* oxidation-reduction transformation strategies [18]. Moreover, nZVI has demonstrated good cleaning percentages for the remediation of trichloroethene, hexavalent chromium, nitrate, lead, cadmium, and DDT [6].

There are various ways to synthesize nZVI, including electrochemical, carbothermal reduction, ultrasound-assisted, and green synthesis. Although nZVI is reactive as a reducing agent, it has limited mobility, poor agglomeration dispersion stability, and difficulties being separated from the remedied soil. The most popular methods involve combining Pd, Pt, Ag, Cu, and Ni with other noble metals to create an alloy to modify the surface and maintain the function of the material. Other techniques include applying a surface coating of synthetic polymers like polyethylene or biopolymers like starch, carboxymethyl cellulose, and guar gum (ethylene glycol). Although if nZVI is integrated on the surface of supports like silica, activated carbon, zeolites, or polymer membranes, the separation of the nanomaterial from the cleansed soil is made easier. In addition, nZVI can be "trapped" using a particle emulsion or dispersion in a biopolymer such as calcium alginate, chitosan, or gum Arabic. Other metal-based nanomaterials include magnetic NPs, SiO_2, Al_2O_3, TiO_2, iron phosphate, goethite, and TiO_2 [19].

CARBON-BASED NANOMATERIALS

Carbonaceous NPs stand out for their distinctive qualities, including their abundant surface area, high microporosity, superior sorption capabilities, and environmentally favorable makeup. Activated carbon NPs, fullerene C60, fullerene C540, SWCNTs, MWCNTs, graphene, and others are all included in some architectural designs [20, 21]. Moreover, carbon-based NPs can be activated or functionalized to provide additional advantages, just like in other environmental remediation applications. CNTs have become more well-known because they have a greater capacity for adsorption than graphene, graphene oxides, biochar, and granular activated carbon. Adsorption is influenced by surface functional groups like -COOH and -OH and the exposure area.

By connecting functional groups like $-NH_2$, -SH, oxidation procedures, coating with nonmagnetic metal oxides, and grafting magnetic iron oxides, the adsorption capacity can be enhanced. Increased surface area, a high surface-to-volume ratio, and consequently high reactivity make materials more prone to flocculation and less suitable for nano remediation. Multi-wall carbon nanotubes have been adequately stabilized using the surfactant poloxamer 407 [21]. Pb^{2+}, Cu^{2+}, Ni^{2+}, and ZN^{2+} can all be removed by CNTs; however, the immobilization of heavy metals depends on pH, the amount of organic matter present, and the presence of silt and clay particles. Moreover, CNTs may remove contaminants from the soil, such as hexachlorocyclohexane, crude oil, Cr (VI), Cd, and heavy metals, while boosting microbial activity and plant growth. The inclusion of CNTs into membrane filtration, separation columns, and an aqueous dispersion are examples of application strategies for CNTs [22].

NANO-PHOTOCATALYSIS IN ENVIRONMENTAL REMEDIATION

Hazardous materials can be converted into products that are suitable for the environment using a photocatalytic process that makes use of extremely reactive hydroxyl and superoxide radicals. The chemical process is accelerated by photocatalysts when there is a lot of sun energy available. Usually, the creation of the valence band and conduction band by the absorption of photons results in a photocatalyst surface redox process. Hole formation in the valence band and electron formation in the conduction band occurs when a photocatalyst is exposed to radiation. The formation of highly reactive and energetic species like hydroxyl (OH-) and superoxide radicals is a result of these photogenerated electron-hole pairs. The reactive species are strong enough to oxidize organic pollutants, degrade waste plastics, eradicate all microorganisms that are waterborne, and degrade air pollutants like NO_2, CO, and NH_3 [23].

After being exposed to light, electron/hole pairs undergo redox reactions, which are the basis of the photochemical process known as Photocatalysis. AOPs like photocatalysts, which completely convert organic molecules into green products and simultaneously cure various pollutants, have shown to be viable alternatives to conventional treatment techniques for minimizing environmental issues. Also, it has been demonstrated that photocatalytic water splitting is a more dependable, sustainable, and safe means of generating molecular hydrogen. Hydrogen combustion only produces water as a byproduct, making it a potentially sustainable alternative to burning fossil fuels. To lessen the burning issue of the energy crisis, cutting-edge research is being done on photocatalytic water splitting, which can produce molecules of oxygen or hydrogen. For molecular hydrogen synthesis or oxygen evolution reactions, a number of photocatalysts have been studied; however, the majority of them have the serious drawback of having low efficiency. As effective photocatalysts for water splitting or hydrogen production, carbon-based photocatalysts, conjugated organic polymers, metal oxides, sulphides, and nitrides are used. However, electron-hole recombination and photodegradation can reduce efficiency. Because of their improved charge separation and light absorption synergism, modified photocatalysts such as bimetallic NPs, hierarchical nanostructures, nanocomposites, oxynitrides, and functionalized NPs might be ideal alternatives for water-splitting processes.

NANO ADSORBENTS AND NANO CATALYSTS FOR WASTEWATER AND SOIL REMEDIATION

Compared to other cleanup methods, nano remediation has shown to be effective at decontaminating groundwater. Several pollutants present in water have been repaired using nanotechnology, including heavy metals, hydrocarbons or other organic compounds, pesticides, and inorganic ions. The use of nano adsorbents and membrane systems based on nanocomposites and nano catalysts is required for *in situ* nano remediation of water, groundwater, and wastewater.

Organic contaminants, heavy metals, and bacteria resistant to antibiotics can all be completely removed from wastewater thanks to the effective adsorption capabilities and good membrane permeability of nanomaterials. Photocatalytic nanomaterials may help maintain the water's natural quality by assisting in the degradation of dangerous algal blooms.

According to reports, Ni-doped $BiVO_4$ nanocomposite can inhibit the development and spread of algal blooms and render algae inactive when exposed to visible light. Malachite green has been efficiently decomposed using co-doped $BiVO_4$, and it has also exhibited effective action in the inactivation of green tides caused by the presence of *Escherichia coli* and *Chlamydomonas pulsatilla* in wastewater. Additionally, the efficacy of $AgeTiO_2$ and MoS_2/Bi_2WO_6 heterostructures for bloom removal and pollutant detoxification in wastewater has been studied. Lindane (hexachlorocyclohexane) is used, a persistent environmental contaminant, has been found to degrade when integrated with nano-biotechnology using Pd/Fe^0 bimetallic NPs [24].

Photocatalysis' usefulness in the disintegration of harmful resistant microorganisms may be explained by the generation of very reactive radicals capable of breaking down these micropollutants present in water. A variety of mechanistic mechanisms for the inactivation and degradation of pollutants in water have been discovered in the literature. ROS, which put the regular functioning of the cell under some stress, are produced as a result of the major step in all of the pathways, which involves the reduction of molecular oxygen (O_2) by the nanomaterials. By causing oxidative damage to the cell's constituent parts, the integrity of the cell and, subsequently, its metabolism is compromised, ultimately resulting in an inactivation or, occasionally, cell death.

The use of carbon nanotubes and nano zerovalent iron (nZVI) in environmental cleanups, such as the removal of pesticides (DDT, Lindane), inorganic anions, and organochlorines from groundwater, has demonstrated an amazing and promising future [25]. Low standard reduction potential, high mobility, high reactivity, and flexibility are the main characteristics that favor nZVI and carbon nanotubes. It is intended for nZVI to undergo Fe^0 to Fe^{2+} and then further oxidative transformation to Fe^{3+} in contaminated water, quickly decreasing both inorganic and organic contaminants.

Due to their capability for adsorption and versatility in functional group attachment, carbon nanotubes are regarded as effective adsorbents. Many studies have investigated the adsorption ability of SWCNT and MWCNT for transforming organics dissolved in wastewater and activated sludge [26]. Carbon nanotubes are used to remove heavy metals, hazardous chemicals, inorganic waste, and volatile organic compounds. Carbon nanotube adsorption activity may be modified further by applying surface modification procedures such as metal ion grafting, acid treatment, and surface impregnation with reactive moieties [27].

When it comes to heavy metal removal from wastewater and pollutant degradation, modified carbon nanotubes with metal oxides like MnO_2, Al_2O_3, and Fe_2O_3 show effective and promising results [28].

COMBINED NANO REMEDIATION

Much study has been conducted recently on combining nano remediation technology with other mitigating techniques. Synergetic research can be defined as the simultaneous use of several nano remediation techniques, as well as their combination with soil flushing or biotreatment. Many nano-cleanup approaches were used concurrently in several studies. Vilardi *et al.* evaluated the effectiveness of nZVI and CNTs for the removal of Cr(VI), selenium (Se), and cobalt (Co) from aqueous solutions in a batch experiment.

The results showed that the primary mechanism for removing Cr(VI) was a reduction, whereas the primary method for removing other metals was adsorption. The results revealed that when nZVI was employed alone without changing the pH, the efficiency of Cr(VI) removal was 100%, while when CNTs-nZVI nanocomposite was utilized, it fell to about 90%, whereas Co and Se had great removal efficiencies utilizing CNTs-nZVI, at 90 and 80%. According to the findings, the CNTs-nZVI nanocomposite had high adsorption effectiveness for cleaning up water contaminated with heavy metals [29]. Zhang *et al.* [30] examined the effectiveness of CMC-stabilized nZVI composited with BC (CMC-nZVI/BC) for the remediation of Cr (VI)-contaminated soil in a distinct experiment. The results indicated that after 21 days, the immobilization efficiency of Cr (VI) was 19.7, 33.3, and 100% when the dose of CMC-nZVI/BC was 11, 27, and 55 g Kg1. The findings imply that adding BC to CMC nZVI might marginally slow the Cr (VI) transition because some CMC nZVI may adsorb to charcoal.

The reduction reaction continued to remediate, resulting in a high Cr total removal efficiency. For the first time, Qian *et al.* [31] examined the effectiveness of biochar-nZVI for the remediation of chlorinated hydrocarbon in the field in a recent study. They employed water pressure-driven packer methods and direct-push methods. The results of the field study showed that the concentrations of chlorinated solvents dropped significantly in the groundwater within 24 hours of the first nZVI injection. Within the following two weeks, the concentrations rose again. The elimination of the chlorinated solvent from groundwater for 42 days was significantly enhanced by using biochar-nZVI, however. The findings imply that biochar-nZVI is a potential combination method for treating groundwater contaminated by chlorinated solvents. Galdames *et al.* [32] created a novel method for removing heavy metals and hydrocarbons from polluted soil that combines nano remediation with bioremediation. The technique specifically blends compost made from organic waste and nZVI.

The findings suggested that even in uncontrolled circumstances, combining compost and nZVI might reduce the number of aliphatic hydrocarbons by up to 60%. Also, they noticed a notable drop in ecotoxicity in the soil biopile. In a different investigation, Alabresm *et al.* [33] investigated using PVP-coated magnetite NPs and bacteria that break down oil to clean up crude oil at the lab scale. According to the findings, within one hour, NPs by themselves eliminated almost 70% of the elevated oil content. Due to the saturation of NPs, the removal efficiency did not, however, improve. On the other hand, after 48 hours, bioremediation caused by bacteria that break down oil, removed 90% of the oil.

Lastly, NPs and microorganisms that break down oil might be combined to eliminate the oil in 48 hours. This was explained by the sorption of oil components to NPs and subsequent bacterial breakdown. When NPs are employed in conjunction with bacteria that degrade oil, more research is required to comprehend the oil removal mechanism.

In a recent long-term field investigation, Czinnerova *et al.* [34] studied how chlorinated ethenes (CEs) degraded utilizing nZVI supported by electrokinetic (EK) treatment (nZVI-EK). EK may boost nZVI migration and lifespan and its effect on soil bacteria. The findings showed a sharp drop in cis-1,2 dichloroethene (cDCE) at about 70, followed by the emergence of new geochemical circumstances due to the appearance of CE's degradation products.

These new circumstances promoted the development of ground and soil bacteria, including bacteria that can breathe organohalides. According to the findings, the nZVI-EK remediation technique is a potentially effective way to remove CE from soil and groundwater while also increasing the availability of bacteria there. In a different study, Boente *et al.* [35] investigated the removal and recovery of toxic metals from polluted soil using a combination of soil washing and nZVI. The findings revealed that Pb, Cu, and Sb had good recovery yields in the magnetically separated fraction, while Hg was concentrated in a nonmagnetic fraction.

Adding nZVI improved the soil cleaning efficiency, notably for a bigger proportion. The findings indicate that the explored methodology makes NPs usable for soil-washing remediation. An activated carbon fiber (ACF)-supported nZVI (ACF-nZVI) composite's potential for removing Cr (VI) from groundwater was examined by Qu *et al.* [36] Also, they carried out a batch experiment to investigate the impact of operating conditions such as nZVI quantity on activated carbon fiber, beginning Cr(VI) concentration, and pH value on the removal of Cr(VI).

The findings suggested that ACF might live within the nZVI aggregation, increasing the nZVI reactivity and Cr(VI) removal efficiency. With increasing Cr(VI) starting concentration, the removal efficiency of Cr(VI) declined, but in an acidic environment, total removal (100% of Cr(VI) was observed in 1 hour of reaction time). The two-stage proposed removal procedure involved physically adsorbing Cr(VI) onto the ACF-nZVI surface area or inner layer in step one and reducing Cr(VI) to Cr(III) in step two. In a different work, Huang *et al.* [37] investigated the effectiveness of the activation of persulfate (PS) utilizing a zeolite-supported nZVI composites (PSZ/nZVI) system. The findings showed that Z/nZVI demonstrated a high capacity for PS activation (1.5 mM) and that TCE was efficiently removed (98.8 at pH 7 within 2 hours). Moreover, the PS-Z/nZVI system demonstrated good TCE efficiency over a broad pH range (4–7).

CURRENT AND FUTURE DEVELOPMENT OF ENVIRONMENTAL NANO APPLICATIONS

Innovative uses of nanotechnology may also come from sectors like medical nanotechnology, which is developing quicker than the previously stated environmental biotechnology applications. Nanostructures can be functionalized in unique ways with the aid of biomolecules. Modern experimental approaches based on actual molecular interactions have verified this approach.

For the construction of heavy-metal removal/recovery membranes, materials, including amyloid proteins and activated porous carbon, were employed. The detrimental consequences of amyloid protein growth in neurons served as an inspiration for these researchers. In order to create amyloid fibrils with cysteine moieties capable of capturing different ions, they modified the tertiary structure of milk proteins. It is essential for this sort of research to locate a cheap supply of biomolecules. Natural proteins also provide a variety of advantages. Due to their considerable combinatorial flexibility in interacting with other molecules and the reasonable manufacturing costs associated with well-established recombinant technologies, they are a suitable alternative for many applications. Additionally, they may include up to 20 different amino acids. Additionally, biotechnology can be used to create environmentally friendly processes for the functionalization of NPs.

To make new cellulose-like polymers functionalized with bespoke moieties, Gao *et al.* recently used microorganisms Komagataeibacter sucrofermentans. As a result, these bacteria are incorporated into the polymer through standard bioreactor methods and glucose monomers with the required chemical changes.

Solvents, stoichiometry, and environmentally hazardous byproducts are omitted from this method. To make it easier to synthesize many cellulose-based NMs with various uses, biotechnological alterations (such as mutagenesis, protein engineering, or gene editing) show considerable promise for improving this biosystem. According to recent breakthroughs, traditional biochemical fungicides could be replaced with RNA-based fungicides. Spraying double-stranded RNAs on plants or fruit induces expression silencing in pathogens by hybridizing with essential mRNAs. However, the short lifespan of naked RNAs in the environment is a big issue. Clay nanosheets shielded the double-stranded RNA and then tested to see whether it might prolong the biocidal action against the fungus. Since its inception, this technique has solely been used to preserve the plant's root system. From silica particles to viral capsules made from plants or recombinant proteins, Chariou and colleagues investigated them all. Throughout the proof-of-concept testing, the biologically produced capsules beat the nematicide in terms of soil penetration and cargo release.

These bio-NPs are genuinely eco-friendly since they decay spontaneously and produce no organic waste. This sector is expected to be significantly impacted by the biofunctionalization of next-generation NPs. A cutting-edge research topic in environmental biotechnology, DNA hybridization allows for the 3D creation of DNA structures. The relatively straightforward yet flexible identification laws between the nucleotides of separate DNA strands can result in a wide range of geometric forms, from simple crossover tiles to polyhedral meshes.

These designs provide a new functionalization toolset, as evidenced by the work of DNA nanorobots that can be loaded with challenging compounds. DNA sheets that interact with thrombin may be thoroughly studied thanks to the information supplied. The thrombin cargo is released, the DNA origami nanotube is opened by the tumor protein nucleolin, and the tumor is coagulated and necrosed as a result. This protein is contained within a nanosheet by additional DNA molecules. Programmability and complexity were shown in the creation of biologically based biorobots. Using biomolecular interactions to generate novel insecticides and remove antibiotic-resistant super-bacteria from the environment is possible. One thing to think about is the price of DNA origami. Biotechnological methods have recently been used to assess the economic feasibility of DNA/RNA chip manufacturing, recombinant bacteria, and naturally existing bacteria capable of exporting RNAs. Soot, ammonia, and triethylamine were chemically encoded by Koman *et al.so* that analytes, such as triethylamine, could be detected in the air .

Photoinduction and ion shifts in the environment may encode sensors' self-precipitation, such as the CarH bacterial transcription factor or plant/fungal LOV domains in biology. It is possible to expand sensors that monitor abiotic particles, such as nano plastics, to track organisms in wastewater facilities or immune-detectable toxins. These jobs are now accomplished by costly and laborious tests such as DNA sequencing and HPLC. Molecular biotechnology has also been proposed to solve large-scale applications like water desalination.

Aquaporin Z, an aquaporin from bacteria, has been used in polymeric membranes to remove salt and provide clean water for over a decade. In order to create porin proteins with superior water solute exclusion capabilities, this process was modified utilizing contemporary protein modeling methods and molecular biology techniques. These proteins might be used to create O_2 biosensors and genetic circuits that could be modified to create useful NMs that respond stoichiometrically to oxygen level changes. Many sensing and purifying options are made possible by DNA molecules' ability to reject complex analytes like proteins. Transmembrane proteins have recently been studied for their potential use in energy harvesting, microarrays, herbicide detection, gas monitoring, and more. Recently, this topic was studied by Ryu *et al.*

Diabetes management is a multi-billion-dollar industry, and these advances are driven by it. However, developing real-time monitoring bionanotechnologies can benefit the safety of people performing hazardous operations in radioactive, possibly poisonous, or confined spaces. Various materials, including patches, tattoos, and wristbands, have been employed as biosensors in healthcare monitoring. For more complicated contaminants or mixtures, enzymes may be used in the setups above [38].

CONCLUSION

There is still a lack of knowledge regarding the synergetic effect of NPs and biotechnologies during a nano-bioremediation process and how these combined technologies respond to contaminants of various types, even though the synergy of NPs and microorganisms for the degradation of some contaminants has been demonstrated in batch experiments. To the greatest of our understanding, there is not any long-term safety data on the usage of NPs with microbes. Bio-NPs have advantages over metallic NPs, such as biodegradability, which means they have a lower environmental impact. Existing nanotechnologies could be utilized to remediate soil, air, or water; however, more cost-effective production methods need to arise. With these materials, the regulatory framework is a key factor. Researchers could be able to improve the knowledge of the interactions of NMs and bio-based technologies throughout remediation operations under varied environmental circumstances, providing justifications for more stringent regulation. Finally, because it provides environmental advantages while being less expensive than other technologies, nano-bioremediation has the ability to improve sustainability significantly. Furthermore, the use of NMs in a variety of applications, in conjunction with biological therapies, has shown high success in the removal of pollutants, opening up new options for tackling environmental concerns.

REFERENCES

[1] L. S. Franqui, L. A. V. De Luna, T. Loret, D. S. T. Martinez, and C. Bussy, "Assessing the adverse effects of two-dimensional materials using cell culture-based models", In: *Nanotechnology Characterization Tools for Environment, Health, and Safety.* Springer, 2019, pp. 1-46.
[http://dx.doi.org/10.1007/978-3-662-59600-5_1]

[2] G. Karthigadevi, S. Manikandan, N. Karmegam, R. Subbaiya, S. Chozhavendhan, B. Ravindran, S.W. Chang, and M.K. Awasthi, "Chemico-nanotreatment methods for the removal of persistent organic pollutants and xenobiotics in water – A review", *Bioresour. Technol.,* vol. 324, p. 124678, 2021.
[http://dx.doi.org/10.1016/j.biortech.2021.124678] [PMID: 33461128]

[3] A. Roy, A. Sharma, S. Yadav, L.T. Jule, and R. Krishnaraj, "Nanomaterials for remediation of environmental pollutants", *Bioinorganic Chemistry and Applications,* 2021.
[http://dx.doi.org/10.1155/2021/1764647]

[4] L. Luo, T. Peng, M. Yuan, H. Sun, S. Dai, and L. Wang, "Preparation of graphite oxide containing different oxygen-containing functional groups and the study of ammonia gas sensitivity", *Sensors (Basel),* vol. 18, no. 11, p. 3745, 2018.
[http://dx.doi.org/10.3390/s18113745] [PMID: 30400230]

[5] X.H. Tai, S.W. Chook, C.W. Lai, K.M. Lee, T.C.K. Yang, S. Chong, and J.C. Juan, "Effective photoreduction of graphene oxide for photodegradation of volatile organic compounds", *RSC Advances,* vol. 9, no. 31, pp. 18076-18086, 2019.
[http://dx.doi.org/10.1039/C9RA01209E] [PMID: 35520578]

[6] F. Guerra, M. Attia, D. Whitehead, and F. Alexis, "Nanotechnology for environmental remediation: Materials and applications", *Molecules,* vol. 23, no. 7, p. 1760, 2018.
[http://dx.doi.org/10.3390/molecules23071760] [PMID: 30021974]

[7] H.Y. Huang, R.T. Yang, D. Chinn, and C.L. Munson, "Amine-grafted MCM-48 and silica xerogel as superior sorbents for acidic gas removal from natural gas", *Ind. Eng. Chem. Res.,* vol. 42, no. 12, pp. 2427-2433, 2003.
[http://dx.doi.org/10.1021/ie020440u]

[8] X. Yang, Z. Shen, B. Zhang, J. Yang, W.X. Hong, Z. Zhuang, and J. Liu, "Silica nanoparticles capture atmospheric lead: Implications in the treatment of environmental heavy metal pollution", *Chemosphere,* vol. 90, no. 2, pp. 653-656, 2013.
[http://dx.doi.org/10.1016/j.chemosphere.2012.09.033] [PMID: 23084516]

[9] T.A. Aragaw, F.M. Bogale, and B.A. Aragaw, "Iron-based nanoparticles in wastewater treatment: A review on synthesis methods, applications, and removal mechanisms", *J. Saudi Chem. Soc.,* vol. 25, no. 8, p. 101280, 2021.
[http://dx.doi.org/10.1016/j.jscs.2021.101280]

[10] F. Mashkoor, and A. Nasar, "Magsorbents: Potential candidates in wastewater treatment technology – A review on the removal of methylene blue dye", *J. Magn. Magn. Mater.,* vol. 500, p. 166408, 2020.
[http://dx.doi.org/10.1016/j.jmmm.2020.166408]

[11] R. Istratie, M. Stoia, C. Păcurariu, and C. Locovei, "Single and simultaneous adsorption of methyl orange and phenol onto magnetic iron oxide/carbon nanocomposites", *Arab. J. Chem.,* vol. 12, no. 8, pp. 3704-3722, 2019.
[http://dx.doi.org/10.1016/j.arabjc.2015.12.012]

[12]   W.M.M. Mahmoud, T. Rastogi, and K. Kümmerer, "Application of titanium dioxide nanoparticles as a photocatalyst for the removal of micropollutants such as pharmaceuticals from water", *Curr. Opin. Green Sustain. Chem.,* vol. 6, pp. 1-10, 2017.
[http://dx.doi.org/10.1016/j.cogsc.2017.04.001]

[13]   T. Mpouras, A. Polydera, D. Dermatas, N. Verdone, and G. Vilardi, "Multi wall carbon nanotubes application for treatment of Cr(VI)-contaminated groundwater; Modeling of batch & column experiments", *Chemosphere,* vol. 269, p. 128749, 2021.
[http://dx.doi.org/10.1016/j.chemosphere.2020.128749]

[14]   D. Lico, D. Vuono, C. Siciliano, J. B Nagy, and P. De Luca, "Removal of unleaded gasoline from water by multi-walled carbon nanotubes", *J. Environ. Manage.,* vol. 237, pp. 636-643, 2019.
[http://dx.doi.org/10.1016/j.jenvman.2019.02.062] [PMID: 30851592]

[15]   S. Roy, S. Manna, S. Sengupta, A. Ganguli, S. Goswami, and P. Das, "Comparative assessment on defluoridation of waste water using chemical and bio-reduced graphene oxide: Batch, thermodynamic, kinetics and optimization using response surface methodology and artificial neural network", *Process Saf. Environ. Prot.,* vol. 111, pp. 221-231, 2017.
[http://dx.doi.org/10.1016/j.psep.2017.07.010]

[16]   N. Delgado, A. Capparelli, A. Navarro, and D. Marino, "Pharmaceutical emerging pollutants removal from water using powdered activated carbon: Study of kinetics and adsorption equilibrium", *J Environ Manage.,* vol. 236, pp. 301-308, 2019.
[http://dx.doi.org/10.1016/j.jenvman.2019.01.116]

[17]   H. Kaur, A. Bansiwal, G. Hippargi, and G.R. Pophali, "Effect of hydrophobicity of pharmaceuticals and personal care products for adsorption on activated carbon: Adsorption isotherms, kinetics and mechanism", *Environ. Sci. Pollut. Res. Int.,* vol. 25, no. 21, pp. 20473-20485, 2018.
[http://dx.doi.org/10.1007/s11356-017-0054-7] [PMID: 28891010]

[18]   P. Cheng, S. Zhang, Q. Wang, X. Feng, S. Zhang, Y. Sun, and F. Wang, "Contribution of nano-zer--valent iron and arbuscular mycorrhizal fungi to phytoremediation of heavy metal-contaminated soil", *Nanomaterials (Basel),* vol. 11, no. 5, p. 1264, 2021.
[http://dx.doi.org/10.3390/nano11051264] [PMID: 34065026]

[19]   M. Stefaniuk, P. Oleszczuk, and Y.S. Ok, "Review on nano zerovalent iron (nZVI): From synthesis to environmental applications", *Chem. Eng. J.,* vol. 287, pp. 618-632, 2016.
[http://dx.doi.org/10.1016/j.cej.2015.11.046]

[20]   L. Marcon, J. Oliveras, and V.F. Puntes, "In situ nanoremediation of soils and groundwaters from the nanoparticle's standpoint: A review", *Sci. Total Environ.,* vol. 791, p. 148324, 2021.
[http://dx.doi.org/10.1016/j.scitotenv.2021.148324] [PMID: 34412401]

[21]   M.P.S.R. Matos, A.A.S. Correia, and M.G. Rasteiro, "Application of carbon nanotubes to immobilize heavy metals in contaminated soils", *J. Nanopart. Res.,* vol. 19, no. 4, p. 126, 2017.
[http://dx.doi.org/10.1007/s11051-017-3830-x]

[22]   M.L. Del Prado-Audelo, I. García Kerdan, L. Escutia-Guadarrama, J.M. Reyna-González, J.J. Magaña, and G. Leyva-Gómez, "Nanoremediation: nanomaterials and nanotechnologies for environmental cleanup", *Front. Environ. Sci.,* vol. 9, p. 793765, 2021.
[http://dx.doi.org/10.3389/fenvs.2021.793765]

[23]   A.S. Ganie, S. Bano, N. Khan, S. Sultana, Z. Rehman, M.M. Rahman, S. Sabir, F. Coulon, and M.Z. Khan, "Nanoremediation technologies for sustainable remediation of contaminated environments: Recent advances and challenges", *Chemosphere,* vol. 275, p. 130065, 2021.
[http://dx.doi.org/10.1016/j.chemosphere.2021.130065] [PMID: 33652279]

[24]   C. Regmi, B. Joshi, S.K. Ray, G. Gyawali, and R.P. Pandey, "Understanding Mechanism of Photocatalytic Microbial Decontamination of Environmental Wastewater", *Front Chem.,* vol. 6, no. February, p. 33, 2018.
[http://dx.doi.org/10.3389/fchem.2018.00033] [PMID: 29541632]

[25]   H.J. Lu, J.K. Wang, S. Ferguson, T. Wang, Y. Bao, and H. Hao, "Mechanism, synthesis and modification of nano zerovalent iron in water treatment", *Nanoscale,* vol. 8, no. 19, pp. 9962-9975, 2016.
[http://dx.doi.org/10.1039/C6NR00740F] [PMID: 27128356]

[26]   V.K. Gupta, I. Tyagi, H. Sadegh, R.S. Ghoshekand, A.S.H. Makhlouf, and B. Maazinejad, "Nanoparticles as Adsorbent; A Positive Approach for Removal of Noxious Metal Ions: A Review", *Science, Technology and Development,* vol. 34, no. 3, pp. 195-214, 2015.
[http://dx.doi.org/10.3923/std.2015.195.214]

[27]   M. Anjum, R. Miandad, M. Waqas, F. Gehany, and M.A. Barakat, "Remediation of wastewater using various nano-materials", *Arab. J. Chem.,* vol. 12, no. 8, pp. 4897-4919, 2019.
[http://dx.doi.org/10.1016/j.arabjc.2016.10.004]

[28]   M. Khajeh, S. Laurent, and K. Dastafkan, "Nanoadsorbents: classification, preparation, and applications (with emphasis on aqueous media)", *Chem. Rev.,* vol. 113, no. 10, pp. 7728-7768, 2013.
[http://dx.doi.org/10.1021/cr400086v] [PMID: 23869773]

[29]   G. Vilardi, T. Mpouras, D. Dermatas, N. Verdone, A. Polydera, and L. Di Palma, "Nanomaterials application for heavy metals recovery from polluted water: The combination of nano zero-valent iron and carbon nanotubes. Competitive adsorption non-linear modeling", *Chemosphere,* vol. 201, pp. 716-729, 2018.
[http://dx.doi.org/10.1016/j.chemosphere.2018.03.032] [PMID: 29547860]

[30]   R. Zhang, N. Zhang, and Z. Fang, "*In situ* remediation of hexavalent chromium contaminated soil by CMC-stabilized nanoscale zero-valent iron composited with biochar", *Water Sci. Technol.,* vol. 77, no. 6, pp. 1622-1631, 2018.
[http://dx.doi.org/10.2166/wst.2018.039] [PMID: 29595164]

[31]   L. Qian, Y. Chen, D. Ouyang, W. Zhang, L. Han, J. Yan, P. Kvapil, and M. Chen, "Field demonstration of enhanced removal of chlorinated solvents in groundwater using biochar-supported nanoscale zero-valent iron", *Sci. Total Environ.,* vol. 698, p. 134215, 2020.
[http://dx.doi.org/10.1016/j.scitotenv.2019.134215] [PMID: 31494413]

[32]   A. Galdames, A. Mendoza, M. Orueta, I.S. de Soto García, M. Sánchez, I. Virto, and J.L. Vilas, "Development of new remediation technologies for contaminated soils based on the application of zero-valent iron nanoparticles and bioremediation with compost", *Resource-Efficient Technologies,* vol. 3, no. 2, pp. 166-176, 2017.
[http://dx.doi.org/10.1016/j.reffit.2017.03.008]

[33]   A. Alabresm, Y.P. Chen, A.W. Decho, and J. Lead, "A novel method for the synergistic remediation of oil-water mixtures using nanoparticles and oil-degrading bacteria", *Sci. Total Environ.,* vol. 630, pp. 1292-1297, 2018.
[http://dx.doi.org/10.1016/j.scitotenv.2018.02.277] [PMID: 29554750]

[34]   M. Czinnerová, O. Vološčuková, K. Marková, A. Ševců, M. Černík, and J. Nosek, "Combining nanoscale zero-valent iron with electrokinetic treatment for remediation of chlorinated ethenes and promoting biodegradation: A long-term field study", *Water Res.,* vol. 175, p. 115692, 2020.
[http://dx.doi.org/10.1016/j.watres.2020.115692] [PMID: 32199189]

[35]   C. Boente, C. Sierra, D. Martínez-Blanco, J.M. Menéndez-Aguado, and J.R. Gallego, "Nanoscale zero-valent iron-assisted soil washing for the removal of potentially toxic elements", *J. Hazard. Mater.,* vol. 350, no. January, pp. 55-65, 2018.
[http://dx.doi.org/10.1016/j.jhazmat.2018.02.016] [PMID: 29448214]

[36]   G. Qu, L. Kou, T. Wang, D. Liang, and S. Hu, "Evaluation of activated carbon fiber supported nanoscale zero-valent iron for chromium (VI) removal from groundwater in a permeable reactive column", *J. Environ. Manage.,* vol. 201, pp. 378-387, 2017.
[http://dx.doi.org/10.1016/j.jenvman.2017.07.010] [PMID: 28697381]

[37]   J. Huang, S. Yi, C. Zheng, and I.M.C. Lo, "Persulfate activation by natural zeolite supported nanoscale

zero-valent iron for trichloroethylene degradation in groundwater", *Sci. Total Environ.,* vol. 684, pp. 351-359, 2019.
[http://dx.doi.org/10.1016/j.scitotenv.2019.05.331] [PMID: 31153081]

[38]   V. Edgar, C. E. Molina-guerrero, and M. Guadalupe, "Use of nanotechnology for the bioremediation of contaminants: a review", *Processes,* vol. 8, no. 7, p. 826, 2020.
[http://dx.doi.org/10.3390/pr8070826]

SUBJECT INDEX

A

Abiotic reductive reactions 183
Acid(s) 3, 9, 10, 12, 21, 22, 41, 71, 73, 74, 75,
  77, 99, 102, 111, 114, 126, 145, 175,
  181, 182
  acrylic 41, 99
  alginic 21
  ascorbic 145
  carboxylic 10, 12, 22, 126
  chlorogenic 22
  citric 181
  dimercaptosuccinic 77
  fulvic 175
  hyaluronic 21
  hydrofluoric 9
  lipoic 111
  nucleic 71, 73, 74, 75, 102
  organic 181
  periodic 114
  stearic 12
Activated carbon fiber (ACF) 189, 190
Activity 57, 125, 143, 141, 147, 157, 187
  carbon nanotube adsorption 187
  cellular phagocytic 143
  enzymatic 125
  inhibited phagocytic 157
  metabolic 147
  phagocytic 141
  photoelectric 57
Adenosine triphosphate 147
Advanced oxidation processes (AOPs) 184,
  186
Agents 24, 25, 174
  antimicrobial 24, 25
  monolithic 174
Agglomeration effects 72
Aggregates 72, 112, 115, 116, 117, 159, 162
  amorphous 112
  ionized nanoparticle 72
Air pollutants 168, 186
Airborne aerosol 141
Alanine aminotransferase 153

Aloe vera 18, 22
Alzheimer's disease 99
Applications 1, 2, 17, 26, 30, 32, 36, 66, 70,
  94, 96, 137, 140, 141, 156, 159, 184,
  190, 193
  biomedical 94, 96, 156
  commercial biotechnological 17
  industrial 159
  solar energy 184

B

Binding 76, 109
  energy 76
  properties 109
Biosynthesis 17, 26
  microbial 26
Biosynthetic 18, 120
  pathway 120
  processes 18
Biotechnology, microbial 1
Boltzmann's constant 128
Bovine serum albumin (BSA) 95, 98, 99, 101,
  115
Brain dysfunctions 169
Brewster angle microscope (BAM) 34
Bronchiolitis 168
Brownian motion 34, 116, 127

C

Camellia sinensis 22
Cancer 82, 84, 141, 146, 148, 149
  colon 82
Carbon nano tubes (CNTs) 157, 158, 159,
  160, 169, 179, 181, 185, 188
Cardiovascular 81, 82
  issues 81
  malfunction 81
  problems 82
Cathodoluminescence 57
  reverses 57
Caveolae pathway 75

www.ingramcontent.com/pod-product-compliance
Lightning Source LLC
Chambersburg PA
CBHW050845220326
41598CB00006B/435